Macrophage-Derived Cell Regulatory Factors

Cytokines

Vol. 1

Series Editor
C. Sorg, Münster

Basel · München · Paris · London · New York · New Delhi · Singapore · Tokyo · Sydney

Macrophage-Derived Cell Regulatory Factors

Volume Editor
C. Sorg, Münster

24 figures and 16 tables, 1989

Basel · München · Paris · London · New York · New Delhi · Singapore · Tokyo · Sydney

Cytokines

Library of Congress Cataloging-in-Publication Data
Macrophage-derived cell regulatory factors.
(Cytokines; v. 1)
Includes bibliographies and index.
1. Macrophages. 2. Cytokines.
I. Sorg, Clemens. II. Series.
[DNLM: 1. Immunity, Cellular. 2. Macrophages – cytology. 3. Macrophages – immunology. 4. Macrophages – physiology. WH 650 M1736]
QR185.8.M3M53 1989 616.07′9 88–32033
ISBN 3–8055–4793–5

Drug Dosage

The authors and the publisher have exerted every effort to ensure that drug selection and dosage set forth in this text are in accord with current recommendations and practice at the time of publication. However, in view of ongoing research, changes in government regulations, and the constant flow of information relating to drug therapy and drug reactions, the reader is urged to check the package insert for each drug for any change in indications and dosage and for added warnings and precautions. This is particularly important when the recommended agent is a new and/or infrequently employed drug.

ISBN 3–8055–4793–5

Contents

Preface

In the early years following the discovery of the macrophage migration inhibitory factor (MIF) in 1966, many more lymphokines were described in rapid sequence and it was the widely held belief that lymphokines are products of activated lymphocytes which act on other lymphoid cells. This belief still prevailed in the ensuing time and led to the suggestion of the term 'interleukin' in 1979. However, already in 1974, S. Cohen and his co-workers could demonstrate that MIF or MIF-like activity is not only released by activated mononuclear leukocytes but also by nonlymphoid cells. He suggested to use the more comprehensive term 'cytokines'.

Considering the development of the lymphokine field in recent years, this term in fact seems to be more appropriate. Lymphokines, originally defined as nonantibody cell regulatory molecules of the immune system, were shown to be by no means restricted to lymphoid cells, neither in their production nor in their action on target cells. The application of lymphokines in clinical therapy has shown that lymphokines can induce many more effects than one would have expected from their action in vitro on lymphoid cells. The concept to view the immune system as an isolated system certainly has its merits as an experimental approach but the time seems to have arrived to generate and evaluate more comprehensive concepts of cytokine function, particularly in view of the complex clinical situations found in resistance to infection, chronic inflammatory diseases or allergies.

It is obvious that the immune response is part of an inflammatory reaction whose overall pathology has been observed and described in detail by pathologists of the late 19th and early 20th century, and whose cellular and molecular mechanisms are still poorly understood. It therefore is our intention to put special emphasis on cytokines that are cell regulatory products derived from and acting on cells participating in an inflammatory response. However, this should not be understood as a limitation as certain cytokines have also been found to act in the regulation of growth and differentiation of normal cells and tissues. No restrictions will also be on the molecular size or chemical nature of cytokines.

The first volume of *Cytokines* contains a rather arbitrary selection of articles on macrophage-derived cell regulatory factors. The intention is to direct spotlights to a few functional aspects of the mononuclear phagocyte system, whose role as an 'accessory' cell to the immune system is certainly not the least one but represents only a small fraction of the functional spectrum of this complex cellular system which figures so prominent in inflammatory reactions.

I wish to thank all the authors for their contributions and their help in getting this new book series started.

Clemens Sorg

Sorg C (ed): Macrophage-Derived Cell Regulatory Factors.
Cytokines. Basel, Karger, 1989, vol 1, pp 1–18

M-CSF: Molecular Cloning, Structure, in vitro and in vivo Functions

Peter Ralph, Mei-Ting Lee, Ilona Nakoinz

Department of Cell Biology, Cetus Corporation, Emeryville, Calif., USA

Introduction

Macrophage colony-stimulating factor (M-CSF or CSF-1) belongs to a group of proteins that promote the production of blood cells [1, 2]. M-CSF is a specific growth and differentiation factor for bone marrow progenitor cells of the mononuclear phagocyte lineage and also stimulates the proliferation of mature macrophages via specific receptors on the responding cells [3, 4]. M-CSF also acts to maintain the viability and function of mature cells and stimulates their effector mechanisms. M-CSF is considerably more complex than other hemopoietic growth factors, IL-1 through IL-6, G-CSF, and GM-CSF. M-CSF is the only dimer, of molecular weights ranging from 33 to 100 kd [2].

Peripheral blood monocytes have a lifetime measured in hours. The bone marrow of the average adult human produces about 2×10^9 monocytes a day, or four pounds of flesh a year. Monocytes and derived tissue macrophages of special types are critically involved in a number of physiologic functions such as recycling iron and other components of effete erythrocytes, trash disposal and remodeling during wound healing and probably embryogenesis and childhood development, phagocytosis and destruction of invading pathogens, initiation of humoral and cellular immune responses by presenting foreign antigens and providing cytokines to B and T lymphocytes. This paper describes the molecular structure and biological functions of the M-CSF system.

Gene, mRNA and Primary Polypeptide Structure of M-CSF

Human Molecular Biology

Partial genomic clones and a cDNA clone for human CSF were first identified by Kawasaki et al. [5]. Subsequent cDNA clones [6, 7] and identification of the intron-exon structure of the gene [7] leads to the following description. M-CSF exists as a single copy gene. The gene is about 21 kb in length, has 10 possible exon regions, and can generate at least 7 sizes of cytoplasmic mRNA from 1.6 to 4.5 kb [7]. M-CSF mRNA is found in a number of human tissues and is predominantly of the largest 4.0–4.5 kb size.

There are at least two different primary polypeptides encoded by the various mRNAs (fig. 1). 'Long' clone coding region specifies a precursor protein form of 522 amino acids plus an N-terminal 32 amino acid signal sequence [6, 7]. 'Short' coding region, found so far only in 1.6 kb mRNA, specifies a precursor of 224 amino acids plus the same 32 residue leader [5, 7]. The amino acid numbering used will refer to residues in the propolypeptide minus the leader sequence. The two polypeptide forms are identical from position 1 to 149, after which the long form has an additional 298 amino acids not found in the short form. The sequence then continues identically for the last 75 residues in the two forms. The extra 298 amino acids arise from the choice of an upstream splicing site in exon 6. Another variant of 438 amino acids may occur through the use of a different exon 6 splice junction during transcription [8]. The other mRNA sizes have been identified or are consistent with differential use of alternate 3′ untranslated exon regions [7, 9]. The frequency of short mRNA is about 1 in 10^6 and of the combined long mRNAs about 1 in 10^5 in the MIA PaCa cell line [5, 7].

Control of M-CSF expression is unknown, but there are TATA-like and CAAT-like sequences upstream of the presumed transcriptional start site and other putative prompter and enhancer regions [9].

Chromosomal Location of the Human M-CSF Gene and Its Receptor

In situ hybridization studies have located the human M-CSF gene on the long arm of chromosome 5 at 5q33.1 [10]. This is very close to the gene for the M-CSF receptor, the protooncogene *c-fms* [11] at 5q33.2-3 [12]. Other interesting genes in this 5q region include GM-CSF and IL-3 only 9,000 bases apart at 5q21-32 [12, 13], IL-5 at 5q31 [14], PDGF and β-adrenergic receptors, an oncogene in the FGF family [15], clotting factor XII [16], and the CD14 monocyte antigen [17] (fig. 2). The $5q^-$ syndrome, characterized by a variety of transpositions or hemizygous deletions in this

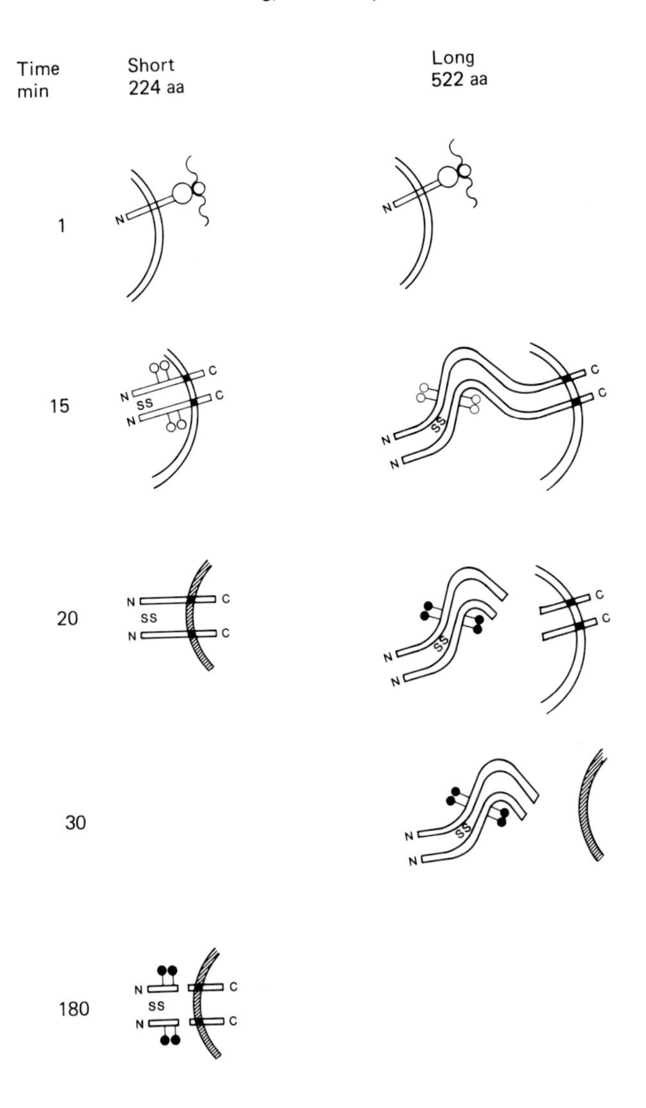

Fig. 1. Processing of short and long form of M-CSF protein. Nascent polypeptides are synthesized on membrane-associated polyribosomes, N-terminally inserted through the endoplasmic reticulum, and modified by N-linked core polysaccharide addition (open circles) and disulfide-linked dimerization (SS). Further N-linked (solid circles) and O-linked (on long form, not shown) polysaccharide modifications in the Golgi and cleavage from the transmembrane region (solid region) at the cell surface (short) or within the secretory apparatus (long M-CSF) is shown.

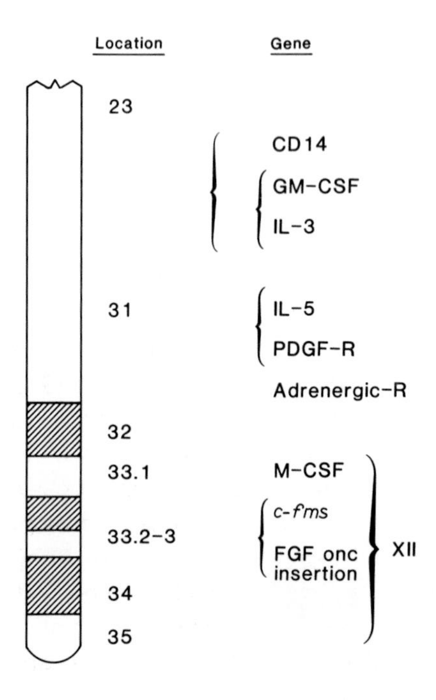

Fig. 2. Location of M-CSF and other genes on human chromosome 5q.

chromosome region, is associated with refractory anemia and secondary myeloid leukemia [12].

Murine Molecular Biology

L cells are the most prolific producers of M-CSF and all the murine cDNA clones have been isolated from this cell line. The complete cDNA clones encode a protein of 520 amino acids plus a 32 residue leader which is very homologous to the human long form [18, 19]. L cells contain mRNA transcripts of 4.5 and 2.3 kb [18, 19]; minor species have also been detected [20]. cDNAs from the two major mRNAs show that they differ only in two positions of coding region (probably not reflected in the secreted, processed protein) and in a different, very long (2 kb) 3′ untranslated region in the longer clone [18, 19]. Transcripts for the short protein form of human M-CSF have not been found so far in the mouse. The cDNA for the murine 4.5 kb mRNA contains 3′ untranslated regions rich in AU sequences [9] which may signal rapid degradation [21, 22]. Despite the production of about 20,000 U/ml M-CSF (0.2 μg/ml) in unstimulated cultures of L cells [23], the frequency of transcripts is less than $1{:}10^6$.

Protein Structure and Cytoplasmic Processing of M-CSF during Synthesis

In both the short and long polypeptide forms of M-CSF, there is a 23 amino acid hydrophobic transmembrane region typical of integral membrane proteins, followed by 36 residues of a short 'cytoplasmic' tail at the C-terminus. Biosynthetic labeling and surface labeling in mouse 3T3 cells expressing the human short cDNA show that M-CSF is rapidly assembled in 68 kd dimers and transported to the cell surface. The soluble hormone is released slowly from the cell surface by proteolytic cleavage at about position 153 for these cells [24] and at position 158 during expression in monkey CV-1 cells [25]. When the human long cDNA is expressed in mouse 3T3 [26] or C129 cells [27] or Chinese hamster ovary cells (CHO) [6], the polypeptide rapidly dimerizes with most of the glycosylated protein N-terminally inserted into the cisternae of the endoplasmic reticulum. It is then rapidly processed and released to the medium in a C-terminal truncated form with monomers of about 223 amino acids. By transient expression studies in monkey COS cells and permanent expression in CHO cells, a cDNA for a monomer of only the first 150 amino acids of M-CSF appears to encode a fully active protein [9, 28].

Evolutionary Conservation of Structure

A comparison of the human and murine M-CSFs is instructive for judging which protein regions may be essential for bioactivity (table 1).

Recombinant proteins containing only the first 150 [9] or 158 amino acids [9, 25] retain full biologic activity with specific activities similar to that of longer forms. The first 7 cysteines are required for biologic activity, in tests of mutating them to alanine or serine; the last two at positions 155 and 161 are not [9]. Nevertheless, all 9 cysteines are conserved between human and mouse.

The regions of strongest homology are the leader peptide, the secreted hormone and the transmembrane sequences, all above 80% identity. Leader peptides have a generally predictable structure [29] and high homology is to

Table 1. Protein homology between human and murine M-CSF

Region	Amino acid position[1]	Percent identity
Leader peptide	−32 to −1	81
Hormone	1–150	81
Hormone	1–223	81
Perimembrane	224–463	56
Transmembrane	464–486	87
Cytoplasmic	487–522	63

Three gaps occur in the murine sequence between positions 246 and 315, and one gap in the human sequence at position 513 to preserve homology.
[1] Numbering of human long clone.

be expected for related, secreted proteins. The transmembrane region consists entirely of hydrophobic amino acids, as is typical of such structures [30]. The retention of this striking feature in both the human and murine genes strongly supports the contention that the complete, membrane-bound propolypeptide has been evolutionarily conserved for an important purpose. This membrane protein form could function as a surface-associated growth factor acting by direct cell-to-cell contact, or it may have a completely different function perhaps even intracellularly.

The high degree of homology of both the minimum hormone length (1–150) and the long protein secreted by many cell types (1 to ca. 223) [P.J. Shadle et al., unpubl. data] suggests that the extra sequence after residue 150 has an important but unrecognized function. The remaining regions, perimembrane (between the N-terminal growth factor and transmembrane regions) and cytoplasmic, have less homology between human and mouse species, 56 and 63%, respectively. This suggests that the structures of these two regions are less critical for biological activity and have been less well conserved. However, a remarkable stretch of 62 amino acids immediately before the transmembrane region shares 81% identity and a subset from position 402 to 434 shows 91% identity. It seems unlikely that all this information is preserved merely to direct intracellular processing to the secreted protein, but an analogous system is the mysterious epidermal growth factor (EGF) gene which specifies about 1,100 amino acids in order to secrete a 53 residue protein [31].

Tissue Expression of M-CSF

Two pancreatic carcinoma cell lines are the strongest known producers of human M-CSF [32]. An SV40-transformed trophoblast line is also a good constitutive producer [6]. Strangely, although fresh human monocytes do not express M-CSF, they can be induced to make their own growth factor, by phorbol myristic acetate (PMA) [33], GM-CSF [34], IFNγ [35] or adherence plus endotoxic lipopolysaccharide (LPS) [36]. M-CSF mRNA and factor secretion have been detected in human alveolar macrophages [37], fibroblasts [38, 39], endothelial cells [40, 41], T lymphocyte lines [6, 42], and probably in B lymphocytes and B lines [43, 44] but not in peritoneal macrophages [37]. M-CSF transcripts have been seen in human liver, placenta [6] and atherosclerotic plaques [A. Wang and M. Ladner, pers. comm.], and in murine liver, lung, brain, heart, spleen and uterus [20]. M-CSF is not detected in mitogen-induced human blood cell or murine spleen cell conditioned media which are rich sources of lymphokines [45, 46]. During pregnancy there is a 10,000-fold increase in M-CSF protein in the mouse uterus which is hormonally regulated. In situ hybridization shows that expression is localized to the luminal and glandular secretory epithelial cells of the endometrium at the sites of, and before, implantation of the developing blastocyst [47]. The juxtaposing trophoblasts of the growing placenta are the only cell type other than mononuclear phagocytes and their progenitors to express M-CSF receptors [11, 48]. M-CSF levels are found in human [S.M. Cox et al., unpubl. data] and murine amniotic fluid [49] and is 2-fold elevated in maternal human serum [50]. This suggests that M-CSF is involved in the growth and maintenance of the placenta during pregnancy.

M-CSF Is a Growth Factor

Human and murine M-CSF stimulate the production of 50–1,000 macrophage cells over 5–7 days of culture from single murine bone marrow progenitors. IL-1 seems to be a cofactor for proliferation of all lineages of hemic cells and is especially active with M-CSF on a subset of early progenitors, each of which can produce up to 0.25 million macrophages [51, 52]. This synergistic effect of IL-1 is accompanied by a 10-fold up-regulation of M-CSF receptors on the stem cells [53]. Murine M-CSF is inactive on human cells and human M-CSF produces only small colonies generally of less than 50 cells, from human bone marrow. The colony cells are large, intensely

Table 2. Regulatory factors in the production of monocytes and macrophages

Stimulatory	References		Inhibitory	References	
	human	murine		human	murine
M-CSF	4,32	1,2,23	TNF-γ	62,63	
IL-1		51,52	TNF-β=LT	63	
IL-3	53	54	IFNs	62,63	
GM-CSF		54	TGF-β		64.65[2]
IL-4		56	Acidic isoferritin	66	66
IL-6		57,58	PGE	67	67
Retinoic acid	59	60	Lactoferrin	68	68
TGF-β	61[1]		Corticosteroids	69	69,70

[1] Stimulates human granulocyte-macrophage colony growth in response to G-CSF + IL-3 + IL-1.
[2] Inhibits murine granulocyte-macrophage colony growth in response to M-CSF.

staining for macrophage nonspecific esterase and well separated from each other, apparently being highly mobile in the agar culture [4, 32]. The concentration of M-CSF required for optimal human marrow cell colony formation is over 10 times that for murine colonies [32], but shows a similar titration in stimulating the function of mature human monocytes and murine macrophages (see below).

Synergistic effects are common among the CSFs, e.g. with M-CSF, GM-CSF, and IL-3 [54]. The production of human macrophage colonies by M-CSF is strongly enhanced by suboptimal amounts of GM-CSF [55]. There are a variety of physiological regulators of monocyte/macrophage production (table 2).

M-CSF Is a Maintenance Factor for Monocytes and Macrophages

Human monocytes are relatively long-lived in culture and take on the appearance of macrophages, but yields can drop by several factors of 2 over a week. M-CSF greatly improves the survival of monocytes, in serum and especially in serum-free conditions [71]. Maintenance of murine peritoneal

Table 3. Stimulating properties of M-CSF on monocytes and macrophages

Function	References	
	human monocyte	murine macrophage
Production of G-CSF	73	74
Production of IFN, TNF and myeloid CSF	75	
Production of prostaglandin E, plasminogen activator, ferritin, IL-1, peroxide		74
Production of procoagulant	76	
Formation of osteoclast-like cells	77[1]	
Tumoricidal activity	79	45
Killing of MAI *(M. avium)*[2]		
Survival, acid phosphatase, surface antigens	71	
Replication of HIV (AIDS virus) together with GM-CSF	80	
Inhibition of vesicular stomatitis virus		81
Intracellular killing of *Candida*		82

[1] In human bone marrow cultures. M-CSF inhibited osteoclast-like cell proliferation in mouse bone marrow culture [78].
[2] In monocytes and macrophages [L.E. Bumudez and L.S. Young, unpubl. data].

and bone-marrow derived macrophages in the presence of M-CSF has been extensively studied by Stanley and co-workers [72].

M-CSF Stimulates Mature Cell Functions

M-CSF in vitro has strong effects on the functions of monocytes and macrophages, including production of cytokines, enzymes, prostaglandins, and reactive oxygen intermediates (table 3). It stimulates nonspecific and antibody-dependent tumorilysis, killing of *Candida* and *Mycobacterium avium*, resistance to viral infection, and suppression of T cell responses. On the other hand, M-CSF together with GM-CSF greatly enhances the replication of human immunodeficiency virus (HIV) in monocytes (table 3). Approximately 1,000 U/ml is optimal for almost all of these effects, as it is for colony formation by human bone marrow [32], but in contrast to 100 U/ml required for murine bone marrow colonies [3].

Table 4. In vivo activation of macrophages and regulation of progenitors by M-CSF[1]

A.	Peritoneal cavity	Total cells		Acid phosphatase
	M-CSF	2.6×10^6		0.139
	Saline	1.6×10^6		0.082
B.	Peritoneal cavity	Total cells		Acid phosphatase
	M-CSF	8.6×10^6		0.172
	Saline	4.2×10^6		0.045
	Spleen	Weight, mg		Cell number
	M-CSF	284		2.6×10^8
	Saline	212		1.9×10^8
C.	Bone marrow	GM colonies per 10^5 cells		
		+media	+M-CSF	+GM-CSF
	M-CSF No.1	0	55	44
	M-CSF No.2	0	46	40
	Saline	0	36	52
	Spleen	Weight, mg		
	M-CSF No.1	286		
	M-CSF No.2	308		
	Saline	228		
	Peritoneal cavity	Total cells		Acid phosphatase
	M-CSF No.1	1.8×10^6		0.186
	M-CSF No.2	2.0×10^6		0.307
	Saline	1.3×10^6		0.206

[1] In experiments A (Lot C36 ARP), B (Lot 100) and C (No. 1 = Lot 104, No. 2 = Lot 105), rhM-CSF (2-5 × 10^7 U/mg) or saline was injected intraperitoneally at 50 μg/dose twice a day for 5 days into female BALB/c mice. Mice were sacrificed 10 days (A) or 4 days (B, C) after the last injection. In D (Lot DCP009), rhM-CSF or human serum albumin was injected intraperitoneally at 75 mg/kg (about 2 mg/dose) twice a day for 6 days, and the mice were sacrificed 16 h later. D mice had enlarged livers, spleens and lymph nodes, especially inguinal. In all cases, marrow from one femur was flushed out, spleens weighed and passed through a screen to produce a single cell preparation, and peritoneal cells harvested. For granulocyte-macrophage (GM) colony formation, spleen cells were puri-

Table 4. (cont.)

D. Bone marrow	Cells harvested/ femur	GM colonies per 10^5 cells			
			+media	+M-CSF	+GM-CSF
M-CSF	4.7×10^6		0	127	128
Albumin	5.2×10^6		0	191	200
Spleen	Weight. mg		GM colonies per 10^6 cells		
M-CSFF	975		0	210	337
Albumin	136		0	7	15
	Cells/spleen	Differential, %			
		large $M\phi$[2]	$M\phi$[3]	Lym	Imm[4]
M-CSF	2.55×10^8	22	10	59	9
Albumin	1.40×10^8	0	7	87	6
Peritoneal cavity	Large cells	Differential, %			
		$M\phi$	Lym	PMN	Basophilic
M-CSF	25×10^6	51	39	10	0
Albumin	7.2×10^6	26	68	0	6
		Antibody-dependent tumorilysis[5]			
		PC6:1	spleen 15:1	6:1	
M-CSF		37%	36%	21%	
Albumin		0%	14%	11%	

fied by Ficoll-Hypaque centrifugation and adherence for 1 h to remove macrophages. Marrow and spleen cells were plated with 100 U/ml M-CSF or GM-CSF (Genzyme, Boston, Mass.) and colonies counted at day 7 [4, 5], means of 5 mice. Production of IL-2 by spleen cells in response to 10 µg/ml concanavalin A and production of superoxide anion by peritoneal macrophages in response to 200 ng/ml PMA did not differ between M-CSF-treated and control mice.

[2] Large vacuolated macrophages.

[3] Morphologically normal monocytes and macrophages.

[4] Immature neutrophils and monocytes with ring nuclei.

[5] Peritoneal macrophages purified by adherence and total (macrophage and K cell effectors) spleen cells tested for killing antibody-coated R1 lymphoma targets at the effector:target ratio shown. R1 is resistant to nonspecific macrophage or NK killing [78].

In vivo Studies in Experimental Animals

Murine and recombinant human M-CSF injections into mice intraperitoneally increase the macrophage content of the peritoneal cavity, induce cycling in marrow progenitors, and greatly increase spleen progenitor cell numbers [83, 84] (table 4). The stimulatory effects of M-CSF on precursors for granulocytes, megakaryocytes and erythrocytes [84] is presumably due to an indirect action in vivo. M-CSF blocks the in vivo myelosuppression caused by injection of interferons [85]. M-CSF induces a 5-fold increase in both blood monocytes and granulocytes and protects mice against a lethal infection with a clinical isolate of *Escherichia coli* [86]. It blocks the growth of an established tumor implanted at a distant site from the route of M-CSF administration [R. Zimmerman and L. Aukerman, pers. commun.]. Macrophages recovered from the peritoneal cavity of M-CSF injected mice have enhanced tumoricidal activity compared to human serum albumin-injected controls (table 4).

Clinical Trials

Clinical trials have been in progress in Japan for several years using partially purified urinary M-CSF. Results show enhanced serum G-CSF activity and monocyte production of G-CSF ex vivo [87], increased blood neutrophil levels in cancer patients who are myelosuppressed by their chemotherapy [88], and increased blood neutrophils in chronic neutropenia of childhood [89]. The eminent clinical trials with recombinant M-CSF should test and extend these results.

The finding of increased neutrophils due to administration of a hormone that acts in vitro only on macrophage precursors can be explained by the induction of G-CSF production in blood monocytes and bone marrow macrophages by exposure to M-CSF [73, 74].

Role of M-CSF in Leukemia

Monocytes and macrophages have abundant numbers of functional receptors for M-CSF [90] and can produce the growth factor (see above). About one-quarter fresh human myeloid leukemic cell samples of several types express M-CSF mRNA or secrete the protein, and a partially overlapping subset expresses M-CSF receptors [91]. Thus, an autocrine proliferation via M-CSF may have a role in the leukemogenesis or progression of some myeloid cancers. The human *c-fms* receptor gene is not oncogenic alone, but when expressed together with its normal ligand, M-CSF transforms mouse 3T3 into tumorigenic cells [92].

No CSF or CSF receptor has been identified in humans in an oncogene state (gross overproduction or altered, permanently signaling receptor). Expression of M-CSF in acute myeloblastic leukemia (AML) cells is correlated with poor growth in culture [93]. G-, GM-, or multi-CSF (IL-3) often stimulates growth of AML in culture, and in 75% of cases addition of M-CSF inhibits growth and induces differentiation to nondividing macrophages [93]. It may therefore be preferable to use M-CSF instead of the other CSFs in clinical trials in myeloid leukemia and preleukemia to restore normal blood granulocyte and monocyte levels.

Acknowledgments

We thank R. Halenbeck, K. Koths, Department of Protein Chemistry and C. Cowgill, Department of Process and Product Development for rhM-CSF, L. Aukerman and R. Zimmerman for M-CSF injected and control mice and advice, and L. Rutkowski for manuscript preparation.

References

1 Metcalf D: The molecular biology and functions of the GM-CSF. Blood 1986;67:257–267.
2 Clark SC, Kamen R: The human hematopoietic colony-stimulating factors. Science 1987:236:1229–1237.
3 Stanley ER, Guilbert LJ, Tushinski RJ, et al: CSF-1, a mononuclear phagocyte lineage-specific hemopoietic growth factor. J Cell Biochem 1983;21:151–159.
4 Das SK, Stanley ER: Structure-function studies of a colony stimulating factor (CSF-1). J Biol Chem 1982;257:13679–13684.
5 Kawasaki ES, Ladner MB, Wang AM, et al: Molecular cloning of a complementary DNA encoding human macrophage-specific colony-stimulating factor (CSF-1). Science 1985;230:291–296.
6 Wong GG, Temple PA, Leary AC, et al: Human CSF-1: molecular cloning and expression of 4 kb cDNA encoding the human urinary protein. Science 1987; 235:1504–1508.
7 Ladner MB, Martin GA, Noble JA, et al: Human CSF-1: gene structure and alternative splicing of mRNA precursors. EMBO J 1987;6:2693–2698.

8 Cerretti DP, Wignall J, Anderson D, et al: Human macrophage colony stimulating factor: alternative RNA and protein processing from a single gene. Mol Immunol 1988;25:761–770.

9 Kawasaki ES, Ladner MB: Molecular biology of macrophage colony-stimulating factor (M-CSF); in Dexter M, Garland JM, Testa NG (eds): Cellular and Molecular Biology of Colony Stimulating Factors. New York, Dekker, 1988, in press.

10 Pattenati MJ, LeBeau MM, Lemons RS, et al: Assignment of CSF-1 to 5q33.1: evidence for clustering of genes regulating haematopoiesis and for their involvement in the deletion of the long arm of chromosome 5 in myeloid disorders. Proc Natl Acad Sci USA 1987;84:2970–2974.

11 Ralph P, Warren MK, Lee M-T, et al: Inducible production of human macrophage growth factor, CSF-1. Blood 1986;68:633–639.

12 LeBeau MM, Epstein ND, Rowley JD: The IL-3 gene is located on human chromosome 5. Proc Natl Acad Sci USA 1987;84:5913–5917.

13 Yang Y-C, Kovacic S, Kriz R, et al: The human genes for GM-CSF and IL-3 are closely linked in tandem on chromosome 5. Blood 1988;71:958–961.

14 Sutherland GR, Baker E, Callen DF, et al: Interleukin-5 is at 5q31 and is deleted in the 5q-syndrome. Blood 1988;71:1150–1152.

15 Delli Bovi P, Basilico C: An oncogene isolated by transfection of Kaposi's sarcoma DNA encodes a growth factor that is a member of FGF family. Cell 1987;50:729–737.

16 Royle NJ, Nigli M, Cool D, et al: Structural gene encoding human factor XII is located at 5q33-qter. Somatic Cell Mol Genet 1988;14:217–221.

17 Goyert SA, Ferrero E, Rettig WJ, et al: The CD14 monocyte differentiation antigen maps to a region encoding growth factors and receptors. Science 1988;239:497–500.

18 DeLamarter JF, Hession C, Semon A, et al: Nucleotide sequence of a cDNA encoding murine CSF-1 (macrophage-CSF). Nucleic Acids Res 1987;15:2389–2390.

19 Ladner MB, Martin GA, Noble JA, et al: cDNA cloning and expression of murine CSF-1 from L929 cells. Proc Natl Acad Sci USA 1988;85:6706–6710.

20 Rajavashisth TB, Eng R, Shadduck RK, et al: Cloning and tissue-specific expression of mouse macrophage colony-stimulating factor mRNA. Proc Natl Acad Sci USA 1987;84:1157–1161.

21 Shaw G, Kamen R: A conserved AU sequence from the 3′ untranslated region of GM-CSF mRNA mediates selective degradation. Cell 1986;46:659–667.

22 Caput D, Beutler B, Hartog K, et al: Identification of a common nucleotide sequence in the 3′-untranslated region of mRNA molecules specifying inflammatory mediators. Proc Natl Acad Sci USA 1986;83:1670–1674.

23 Stanley ER: The macrophage colony stimulating factor, CSF-1. Methods Enzymol 1985;116:564–587.

24 Rettenmier CW, Roussel MF, Ashmun RA: Synthesis of membrane bound CSF-1 in NIH 3T3 cells transformed by cotransfection of the human CSF-1 and c-fms (CSF-1 receptor) genes. Mol Cell Biol 1987;7:2378–2387.

25 Halenbeck R, Shadle P, Lee P-J, et al: Purification of human recombinant CSF-1 from CV-1 monkey cells and generation of a neutralizing antibody. J Biotech, 1988; 8:45–58.

26 Rettenmier CW, Roussel MF: Differential processing of colony-stimulating factor-1 precursors encoded by two human cDNAs. Mol Cell Biol, 1988;8:5026–5034.

27 Manos, MM: Expression and processing of a recombinant human macrophage colony-stimulating factor in mouse cells. Mol Cell Biol, 1988;8:5035–5039.

28 Ralph P, Ladner MB, Wang AM, et al: The molecular and biological properties of the human and murine members of the CSF-1 family; in Webb DR, Pierce CS, Cohen S (eds): Molecular Basis of Lymphokine Action, Clifton Humana Press, 1987, pp 295–311.
29 von Heijne G: Signal sequences – the limits of variation. J Mol Biol 1985;184:99–105.
30 Sabatini DD, Kreibich G, Morimoto T, et al: Mechanisms for the incorporation of proteins in membranes and organelles. J Cell Biol 1982;91:1–22.
31 Doolittle RF, Feng DF, Johnson MS: Computer-based characterization of epidermal growth factor precursor. Nature 1984;307:558.
32 Ralph P, Warren MK, Broxmeyer HE: Inducible production of human CSF-1. Blood 1986;68:633–639.
33 Horiguchi J, Warren MK, Ralph P, et al: Expression of the macrophage specific colony-stimulating factor (CSF-1) during monocytic differentiation. Biochem Biophys Res Commun 1986;141:924–930.
34 Horiguchi J, Warren MK, Kufe D: Expression of the macrophage-specific colony-stimulating factor in human monocytes treated with granulocyte macrophage colony-stimulating factor. Blood 1987;69:1259–1261.
35 Rambaldi A, Young DC, Griffin JD: Expression of the M-CSF (CSF-1) gene by human monocytes. Blood 1987;69:1409–1415.
36 Haskill S, Johnson C, Eierman D, et al: Adherence induces selective mRNA expression of monocyte mediators and proto-oncogenes. J Immunol 1988;140:1690–1694.
37 Becker SJ, Devlin RB, Haskill JS: Differential production of tumor necrosis factor (TNF), macrophage colony stimulating factor (CSF1) and interleukin-1 (IL-1) by human alveolar macrophages. J Immunol, 1988, in press.
38 Fibbe WE, van Damme J, Billiau A, et al: Interleukin-1 induces human marrow stromal cells in long-term culture to produce G-CSF and M-CSF. Blood 1988; 71: 430–435.
39 Fibbe WE, van Damme J: Human fibroblasts produce G-, M- and GM-CSF following stimulation by IL-1 and poly IC. Blood, 1988;72:860–866.
40 Sieff CA, Tsai S, Faller DV: Interleukin-1 induces cultured human endothelial cell production of granulocyte-macrophage colony-stimulating factor. J Clin Invest 1987;79:48–51.
41 Fibbe WE, Daha M: IL-1 and poly IC induce production of G-, M-. GM-CSF by human endothelial cells. Exp Hematol, in press.
42 Takahashi M, Yeong-Man H, Setsuko Y, et al: Macrophage colony-stimulating factor is produced by human T lymphoblastoid cell line, cem-on: identification by amino-terminal amino acid sequence analysis. Biochem Biophys Res Commun 1988; 152:1401–1409.
43 Pistoia V, Ghio R, Roncella S, et al: Production of colony-stimulating activity by normal and neoplastic human B lymphocytes. Blood 1987;69:1340–1347.
44 Reisbach G, Hultner L, Kranz B, et al: Macrophage colony-stimulating activity is produced by three different EBV-transformed lymphoblastoid cell lines. Cellular Immunol 1987;109:246–254.
45 Ralph P, Nakonz I: Stimulation of macrophage tumoricidal activity by CSF-1. Cell Immunol 1987;105:270–279.
46 Nakonz I, Ralph P: Stimulation of macrophage ADCC to tumor targets by lymphokines and rCSF-1. J Leuk Biol 1987;42:353a.

47 Pollard JW, Bartocci A, Arceci R, et al: Apparent role of the macrophage growth factor, CSF-1, in placental development. Nature 1987;330:484–486.
48 Muller R, Slamon DJ, Adamson ED, et al: Transcription of *c-onc* genes *c-ras*^{ki} and *c-fms* during mouse development. Mol Cell Biol 1983;3:1062.
49 Azoulay M, Webb CG, Sachs L: Control of hematopoietic cell growth regulators during mouse fetal development. Mol Cell Biol 1987;7:3361–3364.
50 Bartocci A, Pollard JW, Stanley ER: Regulation of colony-stimulating factor 1 during pregnancy. J Exp Med 1986;164:956–961.
51 Moore MAS, Warren DJ, Souza LM: Synergistic interaction between interleukin-1 and CSFs in hematopoiesis. Recent Adv Leuk Lymphoma 1987;445–446.
52 Mochizuki DY, Tushinsky RJ: IL-1 regulates hematopoietic activity, a role previously ascribed to hemopoietin 1. Proc Natl Acad Sci USA 1987;84:5267–5271.
53 Stanley ER, Bartocci A: Regulation of very primitive multi-potent hemopoietic cells by hemopoietin-1. Cell 1986;45:667–674.
54 Williams DE, Straneva JE, Cooper S, et al: Interactions between purified murine colony-stimulating factors (natural CSF-1, recombinant GM-CSF, and recombinant IL-3) on the in vitro proliferation of purified murine granulocyte-macrophage progenitor cells. Exp Hematol 1987;15:1007–1012.
55 Caracciolo D, Wong GG, Rovera G: rhM-CSF requires subliminal concentrations of GM-CSF for optimal stimulation of human macrophage colony formation in vitro. J Exp Med 1987;166:1851–1860.
56 Rennick D, Yang G: IL-4 (BSF-1) can enhance or antagonize the factor-dependent growth of hemopoietic progenitor cells. Proc Natl Acad Sci USA 1987;84:6889–6892.
57 Wong GG, Witek-Giannotti JS: Stimulation of murine hemopoietic colony formation by human IL-6. J Immunol 1988;140:3040–3044.
58 Ikebuchi K, Wong GG, Clark SC, et al: Interleukin-6 enhancement of interleukin-3-dependent proliferation of multipotential hemopoietic progenitors. Proc Natl Acad Sci USA 1987;84:9035–9040.
59 Dover D, Koeffer HP: Retinoic acid enhances colony stimulating factor induced clonal growth of normal human myeloid progenitor cells in vitro. Exp Cell Res 1982;138:192–199.
60 Goldman R: Enhancement of colony-stimulating factor-dependent clonal growth of murine macrophage progenitors and their phagocytic activity by retinoic acid. J Cell Physiol 1985;123:288–293.
61 Ottmann OG, Pelus LM: Differential proliferative effects of transforming growth factor-β on human hematopoietic progenitor cells. J Immunol 1988;140:2661–2665.
62 Degliantoni G, Murphy M, Kobayashi M, et al: Natural killer (NK) cell-derived hematopoietic colony-inhibiting activity and NK cytotoxic factor: relationship with tumor necrosis factor and synergism with immune interferon. J Exp Med 1985; 162:1512–1517.
63 Broxmeyer HE, Williams DE, Lu L, et al: The suppressive influences of human tumor necrosis factors on bone marrow hematopoietic progenitor cells from normal donors and patients with leukemia: synergism of tumor necrosis factor and interferon-γ. J. Immunol 1986;136:4487–4495.
64 Ohta M, Greenburger JS, Anklesaria P, et al: Two forms of transforming growth factor-β distinguished by multipotential haematopoietic progenitor cells. Nature 1987;329:539.

65 Strassmann G, Cole MD, Newman W: Regulation of colony-stimulating factor 1-dependent macrophage precursor proliferation by type β transforming growth factor. J Immunol 1988;140:2645–2651.
66 Broxmeyer HE, Lu L, Bicknell DC, et al: The influence of purified recombinant human heavy-subunit and light-subunit ferritins on colony formation in vitro by granulocyte-macrophage and erythroid progenitor cells. Blood 1986; 68:1257–1263.
67 Pelus LM: Association between CFU-GM expression of Ia-like (HLA-DR) antigen and control of granulocyte and macrophage productions: a new role for prostaglandin E. J Clin Invest 1982;70:568–573.
68 Broxmeyer HE, Williams DE, Hangoc G, et al: The opposing actions in vivo on murine myelopoiesis of purified preparations of lactoferrin and the colony stimulating factors. Blood Cells 1987;31:13–22.
69 Thompson J, van Furth R: The effect of glucocorticoids on the proliferation and kinetics of promonocytes and monocytes of the bone marrow. J Exp Med 1983; 137:10–24.
70 Metcalf D: Cortisone actions on serum colony-stimulating factor and bone marrow in vitro colony-forming cells. Proc Soc Exp Biol Med 1969;132:391–399.
71 Becker SJ, Warren MK, Haskill S: Colony-stimulating factor-induced monocyte survival and differentiation into macrophages in serum-free cultures. J Immunol 1987;139:3703–3709.
72 Tushinski RJ, Oliver IT, Guilbert LJ, et al: Survival of mononuclear phagocytes depends on a lineage-specific growth factor that the differentiation cells selectively destroy. Cell 1982;28:71–81.
73 Motoyoshi K, Suda T, Kusumoto K, et al: Granulocyte-macrophage colony-stimulating and binding activities of purified human urinary colony-stimulating factor to murine and human bone marrow cells. Blood 1982;60:1378–1386.
74 Ralph P, Warren MK, Ladner MB: Molecular and biological properties of human CSF-1. Cold Spring Harbor Symp Quant Biol 1986;51:679–683.
75 Warren MK, Ralph P: CSF-1 stimulates human monocyte production of interferon, tumor necrosis factor, and myeloid CSF. J Immunol 1986;137:2281–2285.
76 Lyberg T, Stanley ER, Prydz H: Colony-stimulating factor-1 induces thromboplastin activity in murine macrophages and human monocytes. J Cell Physiol 1987; 132:367–376.
77 Van de Wijngaert FP, Tas MC, Van der Meer JWM, et al: Growth of osteoclast precursor-like cells from whole mouse bone marrow: inhibitory effect of CSF-1. Bone Miner 1987;3:97–110.
78 MacDonald BR, Mundy GR, Clark S, et al: Effects of human recombinant CSF-GM and highly purified CSF-1 on the formation of multinucleated cells with osteoclast characteristics in long-term bone marrow cultures. J Bone Miner Res 1986;1:227–233.
79 Sampson-Johannes A, Carlino JA: Enhancement of human monocyte tumoricidal activity by recombinant M-CSF. J Immunol, in press.
80 Gendleman H, Orenstein JM, Martin MA, et al: Efficient isolation and propagation of human immunodeficiency virus on recombinant colony-stimulating factor 1-treated monocytes. J Exp Med 1988;167:1428–1441.
81 Lee M-T, Warren MK: CSF-1-induced resistance to viral infection in murine macrophages. J Immunol 1987;138:3019–3023.

82 Karbassi A, Becker JM, Foster JS, et al: Enhanced killing of *Candida albicans* by murine macrophages treated with macrophage colony-stimulating factor: evidence for augmented expression of mannose receptors. J Immunol 1987;139:417–421.
83 Lotem J, Sachs L: In vivo control of differentiation of myeloid leukemic cells by rGM-CSF and IL-3. Blood 1988;71:375–382.
84 Broxmeyer HE, Williams DE, Hangoc G, et al: Synergistic actions in vivo after administration to mice of combinations of purified natural murine colony-stimulating factor 1, recombinant interleukin-2 and recombinant granulocyte-macrophage colony stimulating factor. Proc Natl Acad Sci USA 1987;84:3871–3875.
85 Koren S, Klimpel GR, Fleischmann WR: Macrophage colony stimulating factor (CSF-1) blocks the myeloid suppressive but not the antiviral or antiproliferative activities of murine alpha, beta, and gamma interferons in vitro. J Biol Response Modif 1986;5:571–580.
86 Chong KT, Langlois L: Enhancing effect of macrophage colony-stimulating factor (MCSF) on leukocytes and host defense in normal and immunosuppressed mice. FASEB J 1988;2:A1474.
87 Ishizaka Y, Motoyoshi K, Hatake K, et al: Mode of action of human urinary colony-stimulating factor. Exp Hematol 1986;14:1–8.
88 Matsumoto K, Kakizoe T, Nakagami Y, et al: Clinical trial of CSF-HU (colony-stimulating factor derived from human urine: P-100) on granulocytopenia induced by anticancer therapy in urogenital cancer patients. Hinyokikia Kiyo (Japan) 1987;33:972–982.
89 Komiyama A: Increases in neutrophil counts by purified human urinary colony-stimularly factor in chronic neutropenia of childhood. Blood 1988;71:41–45.
90 Guilbert LJ, Stanley ER: The interaction of ^{125}I-colony-stimulating factor-1 with bone marrow-derived macrophages. J. Biol Chem 1986;261:4024–4032.
91 Rambaldi A, Wakamiya N, Vellenga E, et al: Expression of the macrophage colony-stimulating factor and c-fms genes in human acute myeloblastic leukemia cells. J Clin Invest 1988;81:1030–1035.
92 Roussel MF, Dull TJ, Rettenmier CW, et al: Transforming potential of the *c-fms* proto-oncogene (CSF-1 receptor). Nature 1986;325:549–551.
93 Miyauchi J, Minden MD, McCulloch EA: The effect of recombinant human CSF-1 on acute myeloblastic leukemia (AML) cells in culture. Leukemia 1988;2:382–387.

Dr. Peter Ralph, Department of Cell Biology, Cetus Corporation,
1400 Fifty-Third Street, Emeryville, CA 94608 (USA)

Sorg C (ed): Macrophage-Derived Cell Regulatory Factors.
Cytokines. Basel, Karger, 1989, vol 1, pp 19–37

Factor Increasing Monocytopoiesis

R. van Furth, W. Sluiter

Department of Infectious Diseases, University Hospital, Leiden, The Netherlands

Introduction

Work done during the last 15 years has shown that the formation of hematopoietic cells in the bone marrow occurs under the influence of humoral factors that are synthesized and secreted both locally in the bone marrow and by cells at distant sites. In 1976, we described a serum factor that induces monocytosis during an acute sterile inflammation [78]. The present contribution, based on research done in mice, gives a review of the origin and kinetics of macrophages, monocytes, and their precursors, a brief summary of some of the characteristics of hematopoietic growth factors, and a detailed discussion of the synthesis, characteristics, and functions of the factor increasing monocytopoiesis (FIM).

Origin and Kinetics of Mononuclear Phagocytes

Pluripotent stem cells residing mainly in the bone marrow are the progenitors of all hematopoietic cells [19, 77]. During normal hematopoiesis, these pluripotent stem cells usually do not undergo division, but can be rapidly mobilized to proliferate when required. It has been postulated that during each division of the pluripotent stem cell, commitment of its progeny is determined by a stochastic process [76]. The committed progeny, which are predominantly in cycle, give rise to the precursors of the various blood-cell lineages, and have only a limited capacity for self-renewal [41]. There is evidence that a bipotent stem cell, called GM-CFU, gives rise to the mononuclear phagocyte and granulocyte cell lines [47, 48]. It is not known whether this bipotent stem cell leads to a stem cell committed to a single cell line or directly to myeloblasts and monoblasts.

The most immature cell that can be recognized as a mononuclear phagocyte is the monoblast, which has a cell-cycle time of about 12 h [31, 32]. Division of a monoblast gives rise to two promonocytes [32]. Promonocytes are the direct precursors of the monocytes [21]. The promonocyte, too, divides only once (cell-cycle time about 16 h) and gives rise to two monocytes [21]. Thus, from monoblast to monocyte there is a fourfold amplification. Monocytes do not divide further, and leave the bone marrow randomly within 24 h after they are formed, which means that they do not pass through a maturation stage in the bone marrow [20]. The total number of monocytes in the bone marrow at any given time is rather small, i.e., for Swiss mice only 2.4×10^6 cells or 1.2% of the total number of bone marrow cells [21]. Monocytes entering the peripheral blood with a rate of 0.62×10^5 monocytes/h [22] are distributed over a circulating pool and a marginating pool, the latter accounting for about 60% of the total of 15.33×10^5 blood monocytes [28]. In the circulation, monocytes remain for a relatively long time (half-time about 17 h) [20] compared with granulocytes (half-time about 7 h). They leave that compartment randomly and then differentiate into macrophages.

In the normal steady state, the maintenance of the population of macrophages in a tissue compartment depends on the influx of monocytes from the circulation and on local division of mononuclear phagocytes that also derive from the bone marrow and divide once in the tissues or body cavities. Calculation has shown that on average 75% or more of the macrophage population is supplied by the influx of monocytes and 25% or less by local division of (immature) mononuclear phagocytes [6, 23, 24, 26, 27]. According to these data, the calculated mean turnover time of the total number of macrophages in the tissues amounts to about 7–14 days [27], which is much shorter than reported in the past [3, 14, 20]. Almost nothing is known about the ultimate fate of the macrophages, except that in all probability macrophages migrate to the local lymph nodes, where they remain or die, and that alveolar macrophages leave the body via air spaces.

Kinetics of Monocytes and Macrophages during Acute Inflammation

During an acute inflammation the number of leukocytes in the blood increases temporarily. In connection with the interest in the kinetics of monocytes and macrophages during inflammation, the choice of inflammatory stimuli used in the experimental models has been aimed at obtaining more monocytes in the circulation and macrophages at the site where the stimulus was applied. For example, a sterile peritonitis induced by newborn calf serum or latex leads to a temporary rise of the number of monocytes [20, 22, 78, 79]. This increase is caused by enhanced production of monocytes in the bone marrow, which is due to an initial decrease of the cell-cycle time of the promonocytes. The number of promonocytes increases as well due to an increase in the mitotic activity of the monoblasts [22]. An inflammatory reaction is initially accompanied by a transient decrease of the mean half-time of circulating monocytes from 17 to 10 h [22].

The increase in the number of macrophages at the site of an acute inflammation is mainly due to an increased influx of monocytes, called exudate macrophages; a relatively small share is taken by local production [4–6, 16, 20, 24, 25, 30].

Hemopoietic Growth Factors

The proliferation and differentiation of hematopoietic cells from pluripotent stem cell to the mature end cell is stimulated and regulated by humoral factors. These factors, in general called hemopoietic growth factors, can be long-range factors produced at distant sites and short-range factors produced in the microenvironment of the proliferating cells.

Hemopoietic growth factors can be divided into multipotential growth factors and lineage-restricted growth factors (fig. 1). The best-known family of growth factors are the colony-stimulating factors (CSF), among which four major types have been identified [49, 50, 56]. Two of these proved to be relatively lineage specific, i.e., granulocyte-CSF (G-CSF) and macrophage-CSF (M-CSF). Multi-CSF, which is also known as interleukin-3 [36], and by other names, stimulates the self-renewal of multipotent stem cells and the proliferation of erythroid, megakaryocytic, eosinophilic, neutrophilic, and macrophage-progenitor cells. Granulocyte/macrophage-CSF (GM-CSF) stimulates the proliferation of neutrophilic granulocytes and macrophages. Some growth factors, e.g., multiCSF and GM-CSF activity, act synergistically on stem cell proliferation [37]. Hemopoietin-1, which is identical to interleukin-1 (IL-1) [53], stimulates the monocyte production indirectly by increasing the number of M-CSF receptors on multipotential stem cells [2], and by stimulation of the production and secretion of M-CSF and GM-CSF [18].

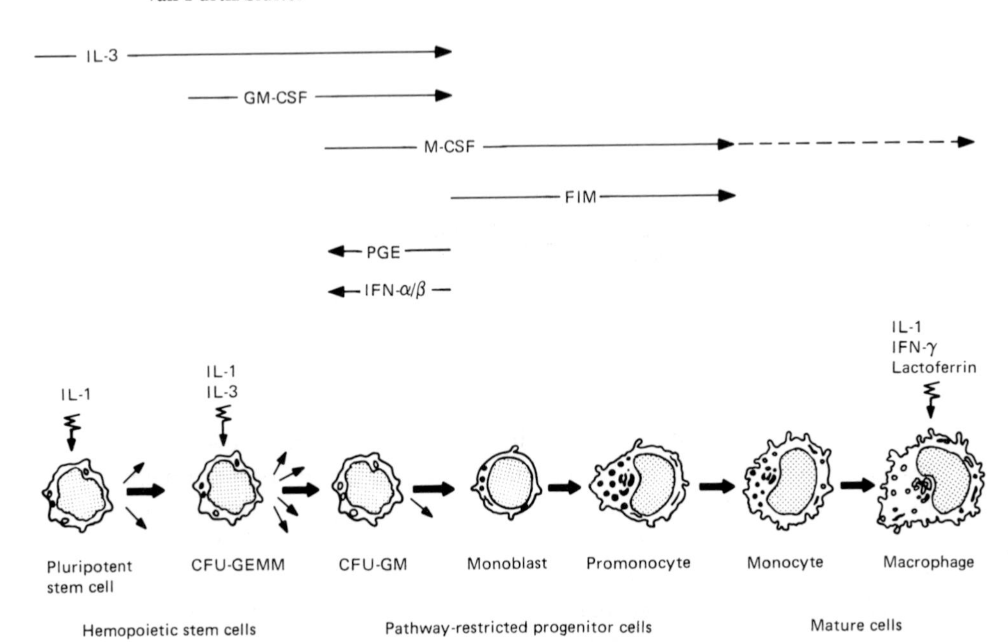

Fig. 1. Schematic representation of factors controlling monocyte production in the bone marrow. Solid lines connect the relevant hematopoietic cells, and thin arrows indicate the descendence of other hematopoetic cell lines, e.g., lymphocytes from pluripotent stem cells. Arrows cover the cells whose proliferation is directly stimulated (→) or inhibited (←) by humoral factors. Factors with only an indirect effect on the proliferation of hematopoietic cells, e.g., by increasing the number of receptors for a growth factor or by stimulating or inhibiting the release of those factors from cells in the mononuclear phagocyte line, are indicated. IL = interleukin; GM-CSF = granulocyte/macrophage colony-stimulating factor; M-CSF = macrophage-CSF; FIM = factor increasing monocytopoiesis; PGE = prostaglandin of the E series; IFN = interferon; CFU-GEMM = colony-forming unit, able to form granulocyte-erythrocyte-monocyte-megakaryocyte progeny; CFU-GM = CFU, able to form granulocytes and macrophages.

CSF can also counterbalance their own stimulatory effect by inducing macrophages to release E-type prostaglandins (PGE) and IFN-α/β [54] and acidic isoferritins [9]. Furthermore, the release of CSF by macrophages is inhibited by lactoferrin from granulocytes [8]. Thus, unlimited stem cell proliferation is avoided by positive and negative feedback control mechanisms involving complex interactions between humoral factors (fig. 1).

Until recently, the role assigned to CSF, PGE, and lactoferrin in the mitotic activity of immature mononuclear phagocytes, including the committed stem cells, was based mainly on in vitro studies. Now, however, recombinant DNA technology permits large-scale production of CSF for use in in vivo studies. Initial studies in mice suggest that the various effects of multi-CSF [10, 39, 42, 51]. GM-CSF [10, 52], and M-CSF [10, 45, 83] in vivo correspond with the effect predicted from their action in vitro.

Humoral Regulation of Monocytopoiesis during Inflammation

During the onset of an inflammatory reaction the peripheral blood contains a factor that stimulates monocytopoiesis in the bone marrow of mice by increasing the rate of division of the monoblasts as well as the number of promonocytes and by reducing the cell-cycle time of the promonocytes [78, 79], changes identical to those seen during acute inflammation [22]. This factor, called factor increasing monocytopoiesis (FIM), has also been found in the blood of rabbits during an acute inflammation [67]. Other investigators have confirmed the presence of a factor that stimulates monocytopoiesis during inflammation [59, 65].

FIM has no effect on the production of granulocytes or lymphocytes and can thus be considered cell-line specific [67, 78, 79], but is not species specific [67]. Characterization showed that FIM is a protein with no carbohydrate moieties essential for its function and having a molecular weight between 18,000 and 25,000. FIM is not a complement component, a clotting factor, or interleukin-1 or CSF-1 (table 1) [70, 80, Sluiter and van Furth, unpubl.].

FIM is synthesized and secreted by macrophages. It is present in extracts of these cells [70, 80], and macrophages secrete this factor during phagocytosis [70]. Alveolar macrophages secrete FIM in vitro in the absence of a phagocytosable particle [71]. The characteristics of FIM secreted by these cells are in all respects similar to those of FIM in serum (table 1).

FIM can only be demonstrated in serum after the induction of an inflammatory reaction, i.e., when a particle that can be phagocytosed is present at the site of inflammation and the number of macrophages in the exudate has risen. Granulocytes and lymphocytes do not contain or secrete FIM [70].

C57BL/10 and CBA mice, which differ in their ability to react to an inflammatory stimulus by an increase in the numbers of circulating monocytes and exudate macrophages, show the same increase of FIM in the blood during the initial phase of an inflammation [68]. C57BL/10 mice, which develop a monocytosis and show increased numbers of exudate macrophages

Table 1. Characteristics of FIM in serum and macrophages

Characteristics	Serum FIM	Macrophage FIM
In vivo stimulation of:		
Monocyte production	yes	yes
Granulocyte production	no	no
Lymphocyte production	no	no
Species specificity	no	not done
Concentration-effect relationship	yes	yes
Chemical nature	protein	protein
Molecular weight	18,000–25,000	10,000–25,000
CSF-1 activity	no	no
IL-1 activity	no	no
Chemotactic activity	no	no

at the site of inflammation, react by an increase of the monocyte production in the bone marrow upon stimulation with FIM from either C57BL/10 or CBA mice. CBA mice, which are low responders to an inflammatory stimulus, do not show increased monocyte production in response to an injection of FIM. These findings constitute evidence that the ability of monocyte precursors to respond to FIM is genetically controlled [68]. Formal proof must, however, await genetic analysis based on studies in F_1 and F_2 hybrid mice or recombinant inbred strains and backcross progeny of the high- and low-responder parental mouse strains.

The mechanism underlying the difference in the response to injected FIM is not fully understood (fig. 2). Monoblasts and promonocytes of low-responder mice might lack binding sites for FIM or these binding sites might be blocked and their monocyte precursors therefore be unable to respond to FIM by increased rate of division. Shielding of binding sites for FIM by an inhibitory factor would be consistent with the observations of Kongshavn et al. [40], who found that the ability of bone marrow chimeras to develop monocytosis during an inflammatory reaction is determined by the recipient. Since the resistance to many kinds of infection is determined by the ability to increase the production of monocytes and the migration of these cells to the site of infection, the genetically controlled sensitivity to FIM might play an important role in the control of such infections.

During the second phase of an inflammatory response the serum contains a factor that inhibits monocytopoiesis [82]. This factor is called the

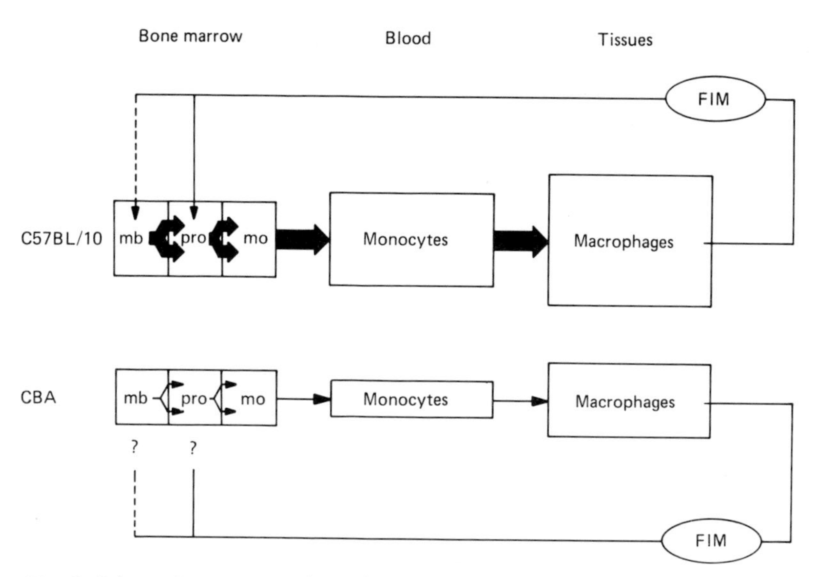

Fig. 2. Schematic representation of the humoral regulation of the production of monocytes during an inflammatory reaction. In both B10 and CBA mice, FIM is produced and secreted by macrophages and is present in the circulation. In B10 mice, FIM stimulates monocyte (mo) production by interaction with promonocytes (pro) and monoblasts (mb); in CBA mice this interaction does not occur and monocyte production is not stimulated.

monocyte production inhibitor (MPI). MPI inhibits the monocytopoiesis in vivo and the proliferation of monoblasts in vitro. Recent investigations [Frendl et al., unpubl. observations] have shown that MPI, whose molecular weight is estimated to be about 250,000, also inhibits the proliferation of macrophage-like cell lines. It remains to be decided whether this factor is identical to the inhibitor that makes bone marrow cells unresponsive to CSF reported to occur in serum [12, 46, 74].

In vivo and in vitro Bioassay for the FIM

At present, two bioassays are available to detect FIM. These assays determine the increase of the number of blood monocytes after intravenous injection of the factor into untreated animals (in vivo method) [67, 68, 78, 79] or establish the increased proliferation rate of a macrophage cell line in vitro (in vitro method) [68, 70].

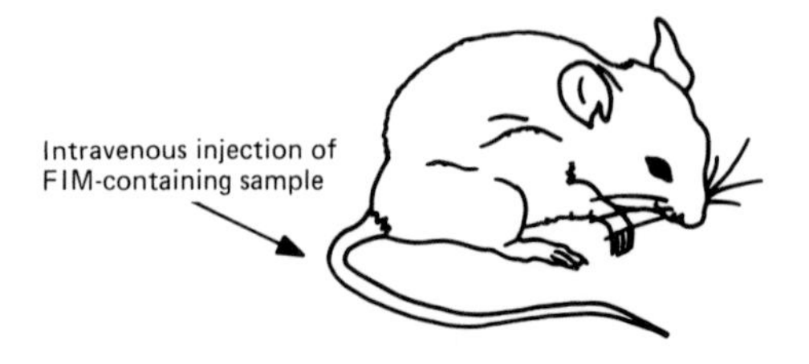

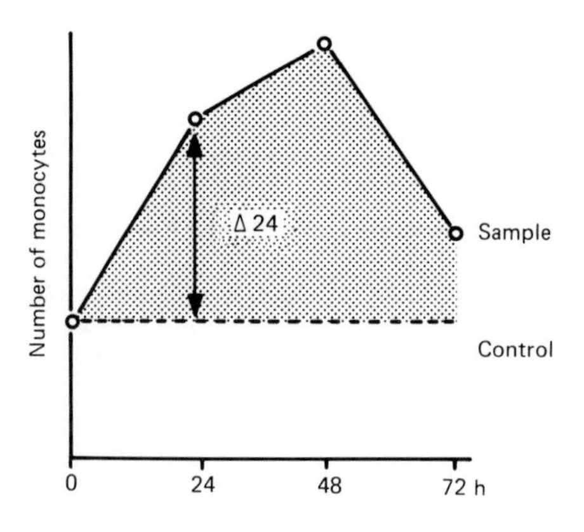

Fig. 3. Schematic representation of the in vivo assay for FIM.

In vivo Bioassay

This assay is performed as follows: normal untreated mice are injected intravenously with serum, serum fractions from blood collected at various times during the inflammatory reaction, or cell-culture supernatant containing FIM, and the number of monocytes in the circulation is counted after various intervals (fig. 3). If these samples contain FIM, the animals can be expected to develop monocytosis. For this purpose, samples of venous blood are taken by cutting off the tip of the tail. However, this procedure causes severe stress [38, 69], and because stress affects the activity of FIM, such mice can only be used once. Small blood samples taken under standard conditions from the retro-orbital plexus cause only minor stress and no granulocytosis or monocytopenia [69]; this is now used routinely.

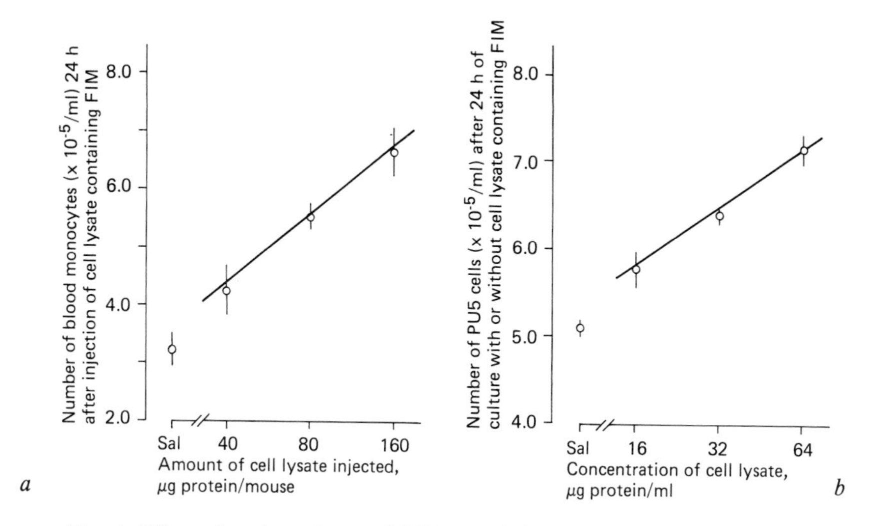

Fig. 4. Effect of various doses of FIM-containing lysates of the cell line J774, expressed as amount of protein, on the number of blood monocytes 24 h after intravenous injection into Swiss mice *(a)* and on the population doubling time of the cell line PU5 *(b)* expressed as the increase in the number of cells after 24 h of culture. The relationships can be described by the following equations:

$N_{mo} = -2.15 + 1.21$ (dose), $R^2 = 0.78$, *(a)*

$N_{pu} = 3.22 + 0.65$ (dose), $R^2 = 0.82$, *(b)*,

where N_{mo} is the number of blood monocytes $\times 10^{-5}$ per milliliter blood, N_{pu} the number of PU5 cells per milliliter after 24 h of culture, and dose the logarithm (base 2) of the amount (μg) of sample per mouse *(a)* or per milliliter *(b)* administered.

The in vivo assay is based on the ability of monocyte precursors to respond to FIM by increased monocyte production, which is measured as the increase in the number of monocytes in the circulation. The reliability of this assay for assessment of the stimulation of monocyte production is evident from the increase in the number of labeled monocytes following simultaneous injection of ^{3}H-thymidine and FIM, which reflects increased mitogenic activity of monocyte precursors [67, 79]. It has also been found that during the first 12 h after the injection of serum containing FIM, the promonocytes almost doubled in number and their cell-cycle time decreased by 25%, a pattern similar to that seen in mice during an acute inflammatory reaction induced by a sterile stimulus [22]. Furthermore, stimulation by FIM is restricted to the production of monocytes, since the numbers of labeled granulocytes and lymphocytes do not change [67, 79]. The effect measured with the in vivo bioassay is dose dependent (fig. 4a), but on a protein basis is less sensitive than the in vitro assay (fig. 4a, b).

An important point is the general health of the mice used in the laboratory and its impact on the in vivo assay of FIM. In biomedical research, increasing use is made of specified pathogen-free (SPF) animals. Although consensus has not been reached as to which pathogens the animals should be free of before they can be considered normal [75], the need for standardization of the microflora and maintenance of the microbiologic status achieved by housing the animals under pathogen-free conditions is obvious [35], particularly for experiments dealing with the regulation of monocyte production [78]. Mice which suffer from a latent infection can be easily identified, because the number of blood monocytes is elevated [81] in samples taken from the retro-orbital plexus [69].

Although FIM is defined by its effect on monocyte production in vivo, ethical and economic reasons make the in vivo bioassay of FIM inappropriate for the large-scale testing required for the biochemical characterization and purification of this factor.

In vitro Assay

The in vitro bioassay for FIM is based on the stimulatory effect of this factor on the population doubling time of two murine macrophage cell lines, i.e., J774.1 (J774) and PU5-1.8 (PU5) [66, 68, 71] (fig. 5). For the assay of FIM, cells from end-log-phase cultures of these cell lines are diluted and recultured in the presence and absence of FIM in Teflon culture bags [44], which are used to avoid adherence. After culture for 24 h, the viable cells are counted and the FIM activity is expressed as percent difference in cell numbers between cultures of the sample under study (N_s) and the reference culture (N_c), according to the formula: $I = (N_s/N_c - 1) \times 100$. This in vitro assay may be considered specific for FIM, since the proliferation of J774 and PU5 cells is not stimulated by a number of other factors regulating hematopoiesis, including M-CSF [13, Sluiter et al., unpubl. observations]. Day-to-day comparison requires that the culture conditions be strictly adhered to, including, for example, not only the presence of serum, which may induce proliferation of quiescent cells [1], but also the surface on which the cells are cultured, e.g., plastic, glass, or Teflon, since the material can influence the recovery of the cells after culture [43, 44]. Furthermore, the sensitivity to the action of growth-regulating factors is determined by the phase of the cell cycle that the cells in culture are in [7, 57, 60]. Therefore, it is necessary to specify whether the cells used at the start of the assay were obtained from stationary or log-phase cultures.

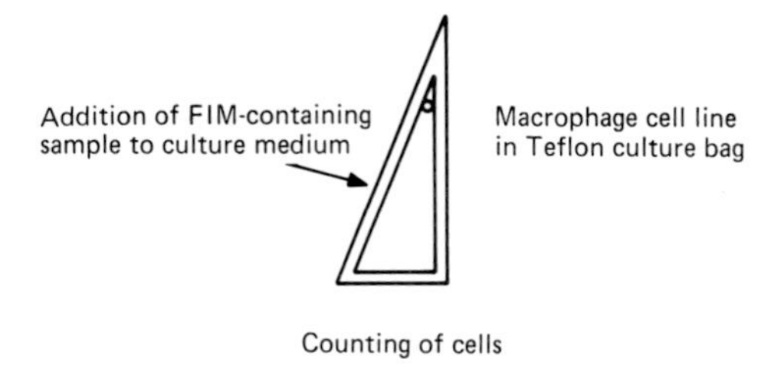

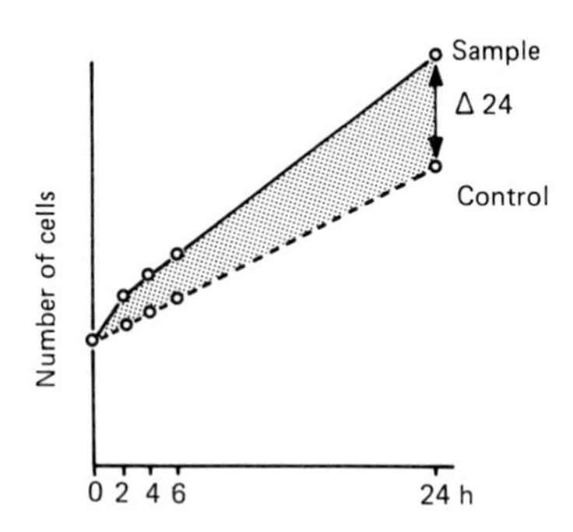

Fig. 5. Schematic representation of the in vitro assay for FIM.

The validity of the in vitro assay is checked by determining the effect of FIM on the cell-cycle time of the macrophage cell line PU5 using the stathmokinetic index [29]. For example, after addition of a FIM-containing serum, the cell-cycle time of PU5 cells was found to decrease by 31.3%. However, the population doubling time decreased by only 18.3% [70]. This means that the effect of FIM on the cell-cycle time was about twice that on the population doubling time, which is the net result of cell growth and cell death. The in vitro assay has a dose-dependent effect (fig. 4b). Furthermore, on a protein basis the in vitro assay is more sensitive than the in vivo assay (fig. 4a).

One potential hazard for cell culture is contamination with mycoplasma species, which can have a drastic influence on the growth, metabolism, and functional activity of mammalian cells [64, 73]. Mycoplasmas can coexist with cell lines for years without producing overt effects [61, 72], and some [63] but not all [58] kinds of macrophage can recover from a mycoplasma infection. Mycoplasma infection of PU5 and J774 cells did not interfere with the stimulatory effect of FIM on the proliferation rate of these cell lines [Sluiter et al., unpubl. observations]. However, because contaminating mycoplasmas can potentially interfere with proper interpretation of the results, uninfected cell cultures are recommended for the in vitro assay of FIM. It should also be kept in mind that cell lines frequently drift away from the normal phenotype and even the genotype over a period of years [55, van Furth and van Schadewijk-Nieuwstad, unpubl. observations], and therefore samples of cultures of the original cell lines should be kept for reference and later use.

Autoregulation of Mononuclear Phagocyte Proliferation by FIM

The synthesis and secretion of FIM by macrophages at the site of an inflammation means that under such conditions the macrophages themselves regulate the supply of monocytes by inducing increased production of these cells in the bone marrow. This in turn leads to an increased number of monocytes in the circulation, and these monocytes then migrate to the site of inflammation, where they interact with the inflammatory stimulus. This process represents a positive feedback mechanism within the mononuclear phagocyte cell line.

An interesting observation has been made in murine macrophage cell lines. Mouse macrophage cell lines proliferate indefinitely in the absence of exogenous growth factors. Extracts of macrophages of various cell lines have FIM activity, and the cells secrete FIM when they ingest opsonized latex beads. Addition of serum containing FIM to a culture medium for cells of a macrophage cell line leads to enhanced proliferation of these cells. This means that the mitotic activity of such cells can be augmented by exogenously supplied FIM, which shortens the cell-cycle time of these cells [25, 29]. When the cell lines were exposed to rabbit anti-FIM IgG antibody, proliferation of the cell line ceased, whereas normal rabbit IgG had no effect [25]. These three findings – i.e., that cells of macrophage cell lines contain and secrete FIM, that the proliferation of macrophage cell lines is augmented by exogenous FIM, and that proliferation of the macrophage cell line is inhibited by anti-FIM antibody – point to some kind of autocrine regulation [25] (fig. 6). In other words, macrophage cell lines produce their own FIM to sustain their proliferation. Other cellular and humoral factors are certainly also involved in the infinite proliferation of these cells.

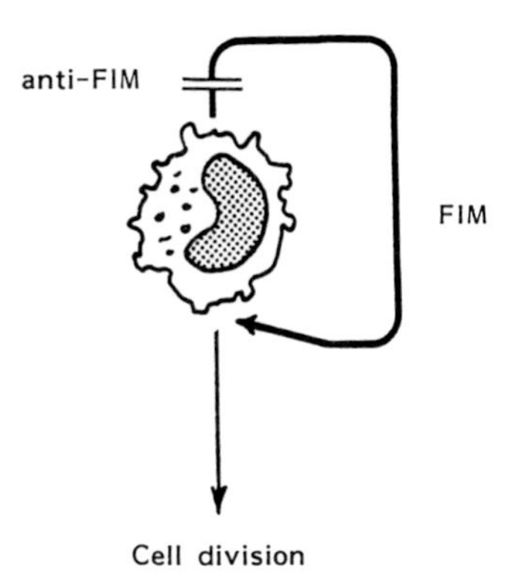

Fig. 6. Schematic representation of the autocrine regulation of the proliferation of macrophage cell lines. Cells of these lines synthesize and secrete FIM; FIM stimulates and anti-FIM antibodies inhibit the proliferation of these macrophages.

Is FIM Distinct from Other Humoral Factors?

Macrophages release a number of factors regulating hematopoiesis, e.g., IL-1, a 15,000 molecular weight protein, CSF and FIM [70] (table 2). We recently found that recombinant human IL-1α, 61% of which is homologous to murine IL-1 at the amino acid level [33], does not have any effect on monocyte production and/or the release of monocytes from the bone marrow [Sluiter et al., unpubl. observations]. However, IL-1 is a potent chemotactic factor for monocytes [17], whereas FIM lacks such activity [80]. We found that IL-1 evokes granulocytosis by inducing the release of these cells from storage pools or their mobilization from the marginating pool, as suggested by Dinarello [17] (table 2). Recently it was conclusively demonstrated that macrophages also release M-CSF [62], a glycoprotein with a molecular weight of 45,000–80.000. M-CSF stimulates the formation of macrophage colonies in the in vitro bone marrow colony assay, but we were unable to detect any effect of M-CSF purified from murine L cells on either the rate of production of monocytes in vivo or the release of monocytes from the bone marrow [Sluiter et al., unpubl. observations] (table 2). Despite the presence of binding sites for M-CSF on the cells of macrophage-like cell lines PU5-1.8 and J774.1 [11, 15, 34], only FIM stimulates the mitotic activity of these macrophage-like cell lines. On the basis of these findings, the conclusion that FIM, IL-1, and M-CSF are different molecules is justified.

Table 2. Characteristics of hematopoiesis-regulating factors produced by macrophages

	FIM	CSF-1 (M-CSF)	IL-1
Specific for mononuclear phagocyte cell line	+	+	–
Stimulation of mitotic rate of monocyte precursors: In vivo	+	?	–
Mobilization into the circulation of:			
Monocytes	+	–	–
Granulocytes	–	–	+
Stimulation of proliferation of macrophage cell lines	+	–	–
Binding sites on bone marrow mononuclear phagocytes, blood monocytes, and macrophage cell lines	?	+	?
Chemotactic action on monocytes	–	?	+
Molecular weight	~20,000	~45,000–80,000	~15,000
Chemical nature	protein	glycoprotein	protein
Site of production	macrophages	macrophages fibroblasts other cells?	macrophages keratinocytes dendritic cells large granular lymphocytes

Does FIM Regulate Monocytopoiesis under Normal Steady-State Conditions?

Peritoneal macrophages exposed to surfactant obtained from normal mice by lung lavage secrete FIM into the culture supernatant [71]. Alveolar macrophages release FIM in the absence of exposure to a phagocytosable material, probably because these cells have ingested surfactant in vivo before explantation [71]. These findings indicate that FIM is secreted in vivo in the lung under steady-state conditions because alveolar macrophages, that continuously ingest surfactant, release FIM into the alveolar space in the absence of an inflammatory stimulus.

The amount of FIM secreted under normal steady-state conditions by alveolar macrophages and macrophages phagocytosing at other sites in the body (e.g., Kupffer cells, spleen macrophages) must be small, because the factor cannot be detected in the serum of normal mice and rabbits with the methods now available [67, 69, 79].

Concluding Remarks

What do the present findings have to say about the human situation in the future? Because of the availability of mass-produced recombinant CSF and IL-1 it may be expected that hemopoietic growth factors will prove to have clinical value for the stimulation of hematopoiesis. The finding that FIM is not specific for mice but also occurs in rabbits, suggests that this factor will be found in other species as well, perhaps including man. If so, it seems likely that FIM too will prove to be clinically useful, for example, to increase host resistance during infection. However, assessment of the therapeutic value of this factor will have to wait until recombinant DNA technology provides sufficiently large amounts of FIM.

References

1 Bar-Shavit R, Kahn AJ, Mann KG, et al: Identification of a thrombin sequence with growth factor activity on macrophages. Proc Natl Acad Sci USA 1986;83:976–980.
2 Bartelmez SH, Stanley ER: Synergism between hemopoietic growth factors (HGFs) detected by their effects on cells bearing receptors for a lineage specific HGF: assay of hemopoietin-1. J Cell Physiol 1985; 122:370–378.
3 Blussé von Oud Alblas A, van Furth R: Origin, kinetics, and characteristics of pulmonary macrophages in the normal steady state. J Exp Med 1979;149:1504–1518.
4 Blussé van Oud Alblas A, van der Linden-Schrever B, Mattie H, et al: The effect of glucocorticosteroids on the kinetics of pulmonary macrophages. J Reticuloendothel Soc 1981;30:1–14.
5 Blussé van Oud Alblas A, van der Linden-Schrever B, van Furth R: Origin and kinetics of pulmonary macrophages during an inflammatory reaction induced by intra-alveolar administration of aerosolized heat-killed BCG. Am Rev Respir Dis 1983;128:276–281.
6 Blussé van Oud Alblas A, Mattie H, van Furth R: A quantitative evaluation of pulmonary macrophage kinetics. Cell Tissue Kinet 1983;16:211–219.
7 Broxmeyer HE, Ralph P: In vitro regulation of a mouse myelomonocytic leukemia line adapted to culture. Cancer Res 1977;37:3578–3584.
8 Broxmeyer HE, Smithyman A, Eger RR, et al: Identification of lactoferrin as the granulocyte-derived inhibitor of colony-stimulating activity production. J Exp Med 1978;148:1052–1067.
9 Broxmeyer HE, Juliano L, Waheed A, et al: Release from mouse macrophages of acidic isoferritins that suppress hematopoietic progenitor cells is induced by purified L cell colony stimulating factor and suppressed by human lactoferrin. J Immunol 1985;135:3224–3231.
10 Broxmeyer HE, Williams DE, Cooper S, et al: Comparative effects in vivo of recombinant murine interleukin-3, natural murine colony-stimulating factor 1, and recombinant murine granulocyte-macrophage colony-stimulating factor on myelopoiesis in mice. J Clin Invest 1987;79:721–730.

11 Byrne PV, Guilbert LJ, Stanley ER: Distribution of cells bearing receptors for a colony-stimulating factor (CSF-1) in murine tissues. J Cell Biol 1981;91:848–853.
12 Chan SH: Influence of serum inhibitors on colony development in vitro by bone marrow cells. Aust J Exp Biol Med Sci 1971;49:553–564.
13 Chen BD-M, Lin H-S: Colony-stimulating factor (CSF-1): its enhancement of plasminogen activator production and inhibition of cell growth in a mouse macrophage cell line. J Immunol 1984;132:2955–2960.
14 Crofton RW, Diesselhoff-den Dulk MMC, van Furth R: The origin, kinetics, and characteristics of the Kupffer cells in the normal steady state. J Exp Med 1978;148: 1–17.
15 Das SK, Guilbert LJ, Forman LW: Discrimination of a colony stimulating factor subclass by a specific receptor on a macrophage cell line. J Cell Physiol 1980; 104:359–366.
16 Diesselhoff-den Dulk MMC, Crofton RW, van Furth R: Origin and kinetics of Kupffer cells during an acute inflammatory response. Immunology 1979;37:7–14.
17 Dinarello CA: Interleukin-1. Rev Infect Dis 1984;6:51–95.
18 Fibbe WE, van Damme J, Billiau A, et al: Interleukin-1 (22-K factor) induces release of granulocyte-macrophage colony-stimulating activity from human mononuclear phagocytes. Blood 1986;68:1316–1321.
19 Ford CE, Hamerton JL, Barnes DW, et al: Cytological identification of radiation chimeras. Nature 1956;177:452–454.
20 van Furth R, Cohn ZA: The origin and kinetics of mononuclear phagocytes. J Exp Med 1968;128:415–435.
21 van Furth R, Diesselhoff-den Dulk MMC: The kinetics of promonocytes and monocytes in the bone marrow. J Exp Med 1970;132:813–828.
22 van Furth R, Diesselhoff-den Dulk MMC, Mattie H: Quantitative study on the production and kinetics of mononuclear phagocytes during an acute inflammatory reaction. J Exp Med 1973;138:1314–1330.
23 van Furth R, Blussé van Oud Alblas A: New aspects on the origin of Kupffer cells; in Knook DL, Wisse E (eds): Sinusoidal Liver Cells. Amsterdam, Elsevier Biomedical, 1982, pp 173–183.
24 van Furth R, Diesselhoff-den Dulk MMC: Dual origin of mouse spleen macrophages. J Exp Med 1984;160:1273–1283.
25 van Furth R, Sluiter W: Macrophages as autoregulators of mononuclear phagocyte proliferation; in Reichard S, Kojima M (eds): Progress in Leukocyte Biology. Macrophage Biology. New York, Liss, 1985, pp 111–123.
26 van Furth R, Nibbering PH, van Dissel JT, et al: The characterization, origin, and kinetics of skin macrophages during inflammation. J Invest Dermatol 1985;85:398–402.
27 van Furth R, Diesselhoff-den Dulk MMC, Sluiter W, et al: New perspectives on the kinetics of mononuclear phagocytes; in van Furth (ed): Mononuclear Phagocytes. Characteristics, Physiology and Function. Dordrecht, Nijhoff, 1985, pp 201–208.
28 van Furth R, Sluiter W: Distribution of blood monocytes between a marginating and a circulating pool. J Exp Med 1986;163:474–479.
29 van Furth R, Elzenga-Claasen I, van Schadewijk-Nieuwstad M, et al: Cell kinetic analysis of a murine macrophage cell line. Eur J Cell Biol 1987;44:93–96.
30 van Furth R, van Zwet TL: Immunocytochemical detection of 5-bromo-2-deoxyuridine incorporation in individual cells. J Immunol Methods 1988;108:45–51.

31 Goud TJLM, Schotte C, van Furth R: Identification and characterization of the monoblast in mononuclear phagocyte colonies grown in vitro. J Exp Med 1975;142:1180–1199.
32 Goud TJLM, van Furth R: Proliferative characteristics of monoblasts grown in vitro. J Exp Med 1975;142:1200–1217.
33 Gubler U, Chua AO, Stern AS, et al: Recombinant human interleukin-1α: purification and biological characterization. J Immunol 1986;136:2492–2497.
34 Guilbert LJ, Stanley ER: Specific interaction of murine colony-stimulating factor with mononuclear phagocytic cells. J Cell Biol 1980;85:153–159.
35 Hanna MG Jr, Nettesheim P, Richter CB, et al: The variable influence of host microflora and intercurrent infections on immunological competence and carcinogenesis. Isr J Med Sci 1973;9:229–238.
36 Ihle JN, Keller J, Henderson L, et al: Procedures for the purification of interleukin-3 to homogeneity. J. Immunol 1982;129:2431–2436.
37 Iscove NN, Roitsch CA, Williams N, et al: Molecules stimulating early red cell, granulocyte, macrophage, and megakaryocyte precursors in culture: similarity in size, hydrophobicity, and charge. J Cell Physiol 1982, suppl 1:65–78.
38 Jyväsjärvi S, Keinänen S, Ruuskanen O: Leucocytosis in the mouse induced by serial blood sampling from the tail. Nature (Lond) 1971;230:122–123.
39 Kindler V, Thorens B, De Kossodo S, et al: Stimulation of hematopoiesis in vivo by recombinant bacterial murine interleukin-3. Proc Natl Acad Sci USA 1986;83:1001–1005.
40 Kongshavn PAL, Sadarangani C, Skamene E: Genetically determined differences in antibacterial activity of macrophages are expressed in the environment in which the macrophage precursors mature. Cell Immunol 1980;53:341–349.
41 Lajtha LG, Pozzi LV, Schofield R, et al: Kinetic properties of haemopoietic stem cells. Cell Tissue Kinet 1969;2:39.
42 Lord BI, Molineux G, Testa NG, et al: The kinetic response of haemopoietic precursor cells, in vivo, to highly purified, recombinant interleukin-3. Lymphokine Res 1986;5:97–104.
43 van der Meer JWM, Bulterman D, van Zwet TL, et al: Teflon film as a substrate for the culture of mononuclear phagocytes. J Exp Med 1978;147:271–276.
44 van der Meer JWM, van de Gevel JS, Elzenga-Claasen I, et al: Suspension cultures of mononuclear phagocytes in the Teflon culture bag. Cell Immunol 1979;42:208–212.
45 Metcalf D, Stanley ER: Haematological effects in mice of partially purified colony stimulating factor (CSF) prepared from human urine. Br J Haematol 1971;21:481–492.
46 Metcalf D, Russell S: Inhibition by mouse serum of hemopoietic colony formation in vitro. Exp Hematol 1976;4:339–353.
47 Metcalf D, Burgess AW: Clonal analysis of progenitor cells commitment of granulocyte or macrophage production. J Cell Physiol 1982;111:275–283.
48 Metcalf D, Merchav S, Wagemaker G: Commitment by GM-CSF or M-CSF of bipotential GM progenitor cells to granulocyte or macrophage formation; in Baum SJ, Ledley GD, Thierfelder S (eds): Experimental Hematology Today. New York, Karger, 1982, pp. 3–9.
49 Metcalf D: The granulocyte-macrophage colony-stimulating factors. Science 1985;229:16–22.
50 Metcalf D: The molecular biology and functions of the granulocyte-macrophage colony-stimulating factors. Blood 1986;67:257–267.

51 Metcalf D, Begley CG, Johnson GR, et al: Effects of purified bacterially synthesized murine multi-CSF (IL-3) on hematopoiesis in normal adult mice. Blood 1986;68:46–57.
52 Metcalf D, Begley CG, Williamson DJ, et al: Hemopoietic responses in mice injected with purified recombinant murine GM-CSF. Exp Hematol 1987;15:1–9.
53 Moore MAS, Warren DJ: Synergy of interleukin-1 and granulocyte colony-stimulating factor: In vivo stimulation of stem-cell recovery and hematopoietic regeneration following 5-fluorouracil treatment of mice. Proc Natl Acad Sci USA 1987;84:7134–7138.
54 Moore RN, Pitruzello FJ, Larsen HS, et al: Feedback regulation of colony-stimulating factor (CSF-1)-induced macrophage proliferation by endogenous E prostaglandins and interferon-α/β. J Immunol 1984;133:541–543.
55 Nibbering PH, van Furth R: Quantitative immunocytochemical characterization of four murine macrophage-like cell lines. Immunobiology 1988;176:432–439.
56 Nicola NA, Vadas M: Hemopoietic colony-stimulating factors. Immunol Today 1984;5:76–80.
57 Pluznik DH, Cunningham RE, Noguchi PD: Colony-stimulating factor (CSF) controls proliferation of CSF-dependent cells by acting during the G1 phase of the cell cycle. Proc Natl Acad Sci USA 1984;81:7451–7455.
58 Polak-Vogelzang AA, Brugman J, Osterhaus ADME, et al: Elimination of mycoplasmas from cell cultures by means of specific bovine antiserum. Zentralbl Bakteriol Hyg A 1987;264:84–92.
59 Punjabi CJ, Galsworthy SB, Kongshavn PAL: Cytokinetics of mononuclear phagocyte response to listeriosis in genetically-determined sensitive and resistant murine hosts. Clin Invest Med 1984;7:165–172.
60 Ralph P, Broxmeyer HE, Nakoinz I: Immunostimulators induce granulocyte/macrophage colony-stimulating activity and block proliferation in a monocyte tumor cell line. J Exp Med 1977;146:611–616.
61 Ralph P: Continuous cell lines with properties of mononuclear phagocytes; in Adams DO, Edelson PJ, Koren HS (eds): Methods for Studying Mononuclear Phagocytes. New York, Academic Press, 1981, pp 155–173.
62 Rambaldi A, Young DC, Griffin JD: Expression of the M-CSF (CSF-1) gene by human monocytes. Blood 1987;69:1409–1413.
63 Schimmelpfeng I, Langenberg U, Hinrich Peters J: Macrophages overcome mycoplasma infections of cells in vitro. Nature 1980;285:661–662.
64 Shin S-I, van Diggelen OP: Phenotypic alterations in mammalian cell lines after mycoplasma infection; in McGarrity GL, Murphy DG, Nichols WW (eds): Mycoplasma Infected Cell Cultures, New York, Plenum Press, 1978, pp 191–212.
65 Shum DT, Galsworthy SB: Stimulation of monocyte production by an endogenous mediator induced by a component from *Listeria monocytogenes*. Immunology 1982;46:343–351.
66 Sluiter W, van Waarde D, Hulsing-Hesselink E, et al: Humoral control of monocyte production during inflammation; in van Furth R (ed): Mononuclear Phagocytes. Functional Aspects. The Hague, Nijhoff, 1980, pp. 325–339.
67 Sluiter W, Elzenga-Claasen I, Hulsing-Hesselink E, et al: Presence of the factor increasing monocytopoiesis (FIM) in rabbit peripheral blood during an acute inflammation. J Reticuloendothel Soc 1983;34:235–252.
68 Sluiter W, Elzenga-Claasen I, van der Voort van der Kleij-van Andel A, et al: Differences in the response of inbred mouse strains to the factor increasing monocytopoiesis. J Exp Med 1984;159:524–536.

69 Sluiter W, Hulsing-Hesselink E, Elzenga-Claasen I, et al: Method to select mice in the steady state for biological studies. J. Immunol Methods 1985;760:135–143.
70 Sluiter W, Hulsing-Hesselink E, Elzenga-Claasen I, et al: Macrophages as origin of factor increasing monocytopoiesis. J Exp Med 1987;166:909–922.
71 Sluiter W, van Hemsbergen-Oomens LWM, Elzenga-Claasen I, et al: Effect of lung surfactant on the release of factor increasing monocytopoiesis by macrophages. Exp Hematol 1988;16:93–96.
72 Stanbridge EJ, Doersen C-J: Some effects that mycoplasmas have upon their infected host; in McGarrity GJ, Murphy DG, Nichols WW (eds): Mycoplasma infection of cell cultures. New York, Plenum Press, 1978, pp 119–134.
73 Stanbridge EJ: Mycoplasma detection – an obligation of scientific accuracy. Isr J Med Sci 1981;17:563–568.
74 Stanley ER, Robinson WA, Ada GL: Properties of the colony stimulating factor in leukaemic and normal mouse serum. Aust J Exp Biol Med Sci 1968;46:715–726.
75 Syed SA, Abrams GD, Freter R: Efficiency of various intestinal bacteria in assuming normal functions of enteric flora after association with germ-free mice. Infect Immun 1970;2:376–386.
76 Till JE, McCulloch EA, Siminovitch L: A stochastic model of stem cell proliferation, based on the growth of spleen colony-forming cells. Proc Natl Acad Sci USA 1964;51:29–36.
77 Vos O, Davids JAG, Weyzen WWH, et al: Evidence for the cellular hypothesis in radiation protection by mouse bone marrow cells. Acta Physiol Pharmacol Neerl 1956;4:482–486.
78 van Waarde D, Hulsing-Hesselink E, van Furth R: A serum factor inducing monocytosis during an acute inflammatory reaction caused by newborn calf serum. Cell Tissue Kinet 1976;9:51–63.
79 van Waarde D, Hulsing-Hesselink E, Sandkuyl LA, et al: Humoral regulation of monocytopoiesis during the early phase of an acute inflammatory reaction caused by particulate substances. Blood 1977;50:141–153.
80 van Waarde D, Hulsing-Hesselink E, van Furth R: Properties of a factor increasing monocytopoiesis (FIM) occurring in the serum during the early phase of an inflammatory reaction. Blood 1977;50:727–741.
81 van Waarde D, Bakker S, van Vliet J, et al: The number of monocytes in mice as a reflection of their condition and capacity to react to an inflammatory stimulus. J Reticuloendothel Soc 1978;24:197–204.
82 van Waarde D, Hulsing-Hesselink E, van Furth R: Humoral control of monocytopoiesis by an activator and an inhibitor. Agents Actions 1978;8:432–437.
83 Yanai N, Yamada M, Wanatabe Y, et al: The granulopoietic effect of human urinary colony stimulating factor on normal and cyclophosphamide treated mice. Exp Hematol 1983;11:1027–1036.

Dr. R. van Furth, Department of Infectious Diseases, University Hospital,
Building 1, C5-P, PO Box 9600, NL-2300 RC Leiden (The Netherlands)

Sorg C (ed): Macrophage-Derived Cell Regulatory Factors.
Cytokines. Basel, Karger, 1989, vol 1, pp 38–53

Effects of Immune Cell Products on Bone

Gregory R. Mundy, Lynda F. Bonewald

Department of Medicine, Division of Endocrinology and Metabolism,
University of Texas Health Science Center, San Antonio, Tex., USA

In 1972, a biological activity was described in the media harvested from activated leukocyte culture supernatants which stimulated osteoclast activity [1]. This activity was detected by its capacity to stimulate organ cultures of rodent fetal or newborn bones to resorb and release previously incorporated ^{45}Ca or matrix constituents. Since the histology of the cultured bones after 48–72 h of culture showed intense increase in cellular activity, and particularly an accumulation of many large multinucleated osteoclasts, the activity was referred to as osteoclast activating factor (OAF). It was produced by leukocytes stimulated with mitogens such as phytohemagglutinin, pokeweed mitogen, concanavalin A or with antigens such as PPD [1, 2]. There was considerable interest in this activity because such an action would account for the bone destruction which occurred adjacent to collections of chronic inflammatory cells in diseases such as rheumatoid arthritis and periodontal disease and similar biological activity was detected in media harvested from cultures of human myeloma cells [3], a B-cell lymphoproliferative disorder.

In the ensuing years, it has become apparent that immune cells under the appropriate stimuli release a myriad of cytokines which influence osteoclastic bone resorption. Many, but not all, of these factors are peptides. They include interleukin-1, tumor necrosis factor, lymphotoxin, 1,25-dihydroxyvitamin D, prostaglandins of the E series, gamma interferon, colony-stimulating factors and transforming growth factor beta. Some act as stimulators and some as inhibitors of osteoclast activity. In addition, their sites of action vary, some acting on osteoclast progenitor proliferation, some on differentiation of committed progenitors and some on preformed mature cells. In some cases, their effects on osteoclasts are indirect. It is the purpose of this review to describe current knowledge of the effects of these immune cell products on osteoclastic bone resorption. Characterized factors which modulate bone cell function will be described individually.

Interleukin-1

Interleukin-1 (IL-1) is a monocyte product which has been found to have numerous biological activities but which has been characterized mainly by its capacity to stimulate thymocyte proliferation. It has been summarized in another chapter in this volume. It has been known for a number of years that it is both a potent as well as a powerful stimulator of bone resorption in vitro. Gowen et al. [4] demonstrated that highly purified human IL-1 preparations resorb organ cultures of mouse calvaria in vitro. Later, it was found that recombinant human IL-1 also stimulates bone resorption in organ culture [5]. The effects of IL-1 on bone are very potent. It stimulates bone resorption at concentrations which are several orders of magnitude lower than any previously described bone resorbing protein. Purified porcine IL-1 has also been found to have similar bone resorbing activity [6].

The mechanisms by which IL-1 stimulates osteoclastic bone resorption have recently been examined. Thomson et al. [7] found that IL-1 did not stimulate mature disaggregated osteoclasts directly to form a resorption lacunae. Bone resorption was stimulated by IL-1 only when the osteoclasts were cocultured with osteoblastic cells. This effect was independent of prostaglandin synthesis since indomethacin did not inhibit it. However, IL-1 stimulates bone resorption not just by a direct effect on mature osteoclasts. In marrow culture systems, it also causes formation of multinucleated cells with osteoclast characteristics from mononuclear precursors in concentrations as low as 2.5×10^{-13} *M*. Again, the prostaglandin synthesis inhibitor indomethacin has no effect on the formation of cells with osteoclast characteristics. Thus, IL-1 may stimulate osteoclastic bone resorption by (1) increasing the activity of mature osteoclasts and (2) increasing the formation of osteoclasts by stimulating proliferation and fusion of osteoclast precursors.

The effects of IL-1 on bone resorption and calcium homeostasis have also been investigated in vivo. Recombinant human IL-1 alpha and beta were infused into intact normal mice [8]. The cytokines were infused for 72 h subcutaneously and caused a marked dose-dependent increase in the plasma calcium. Bone sections were assessed by quantitative histomorphometry and showed evidence of increased osteoclast numbers as well as bone resorption surfaces. Although the effects were similar to those seen with infusions of parathyroid hormone, the effects on bone were found with lower concentrations. When IL-1 was infused at higher concentrations (which caused some animals to die), the animals became hypocalcemic and markedly hyperphosphatemic immediately before death. The mode of death was not certain but may be related to renal failure and the preterminal hypocalcemia probably occurred as a result of the increase in plasma phosphate associated with impaired glomerular filtration.

There has been some debate over whether the effects of IL-1 on bone could be mediated directly or indirectly through prostaglandin synthesis. In some systems, and particularly in mouse bone culture systems, prostaglandin generation has been found to occur in response to IL-1 treatment. However, in these particular systems, the bone cultures produce prostaglandins spontaneously immediately after explantation and the data are difficult to interpret. When IL-1 is infused or injected in vivo, part but not all of the bone resorbing response may be inhibited by the prostaglandin synthesis inhibitor indomethacin [9]. The relationship between cytokines and prostaglandin synthesis remains uncertain. There may be important species differences. However, it is clear that the inhibition of prostaglandin synthesis will not impair all of the bone resorption induced by cytokines such as IL-1.

IL-1 also has effects on bone formation both in vitro and in vivo. When bones are stimulated to resorb with injections of IL-1 in vivo, there will be subsequent replacement of the resorbed bone by new bone formation [9]. The acute effects of IL-1 on bone cells in vitro are to inhibit collagen synthesis [10, 11]. When bone collagen content is measured by the incorporation of proline into collagen and noncollagen protein in 20-day-old fetal rat calvariae incubated with recombinant IL-1, decreased bone collagen synthesis as well as noncollagen protein synthesis occurs. IL-1 stimulates DNA synthesis in the cultured bones. However, when IL-1 is used in much smaller concentrations for short periods, there is a transient stimulation of bone collagen synthesis, which appears dependent on prostaglandin synthesis since it may be inhibited by indomethacin [10].

IL-1 may be involved in pathologic increases in bone resorption. One group has suggested that increased constitutive production of IL-1 by peripheral blood monocytes occurs in some patients with osteoporosis [12]. There are also several reports suggesting it may be produced by tumor cells in vitro [13, 14]. We have found that it can clearly disrupt calcium homeostasis [8]. It is also likely to be produced by chronic inflammatory cells occurring adjacent to bone surfaces in diseases such as rheumatoid arthritis and periodontal disease.

Tumor Necrosis Factors (Lymphotoxin and Tumor Necrosis Factor)

Both the lymphokine lymphotoxin (LT) and the monokine tumor necrosis factor (TNF) stimulate bone resorption in organ culture. LT and TNF are two multifunctional cytokines which have similar cytolytic and cytostatic effects on neoplastic cell lines and appear to mediate their effects through the same receptor. They have only partial sequence homology at the amino acid level and are encoded by separate single copy genes. Recombinant human TNF alpha and LT both stimulate osteoclastic bone resorption when incubated with fetal rat long bones or neonatal mouse calvariae [15, 16]. The bone resorbing activity is dose dependent from 10^{-7} to 10^{-11} *M*. The time-course of action of the cytokines is similar to that for IL-1, although in these organ culture systems they appear to be less potent than IL-1. Their effects on organ cultures of bone are similar to those of parathyroid hormone and can be inhibited by calcitonin, a specific inhibitor of osteoclast activity. Both TNF and LT inhibit bone collagen synthesis when incubated with fetal rat calvariae in vitro at similar concentrations [11, 15]. However, they stimulate DNA synthesis in these cultures [11].

Like IL-1, TNF and LT seem to mediate their effects on preformed osteoclasts indirectly by stimulating cells in the osteoblast lineage to secrete a second factor which activates osteoclasts to form resorption lacunae [17]. In addition to this effect, they also stimulate proliferation and differentiation of osteoclast precursors using the modified Dexter marrow culture [18].

Both TNF and LT cause an increase in plasma calcium when infused or injected in vivo [16, 19]. These factors have clear pathologic significance. Human myeloma cells secrete LT and it appears to be the major mediator of bone destruction in this disease [19]. Myeloma is a neoplastic disorder characterized by extensive osteolytic bone lesions which are accompanied by severe pain and susceptibility to fracture following severe injury. Hypercalcemia occurs in approximately 20–40% of patients sometime during the course of the disease. In this disorder, localized increases in osteoclast activity occur adjacent to the myeloma cells and are responsible for the bone destruction [3]. Cultured myeloma cells produce an osteoclast-stimulating factor which is similar to the bone-resorbing activity produced by activated leukocytes [3, 20].

Tumor cell lines derived from patients with myeloma which are accompanied by osteolytic bone lesions express both LT and TNF messenger RNA. However, only LT biological activity can be detected in the media bathing the cells, and most of the bone-resorbing activity produced by the cultured myeloma cells can be inhibited by means of specific LT monoclonal antibodies [19].

TNF is responsible for several paraneoplastic syndromes including suppression of erythropoiesis and wasting or cachexia. It is also one of the mediators of endotoxic shock. Since it appears to be produced excessively in some patients with neoplastic disease, its excess production by normal immune cells in patients with cancer may also lead to bone destruction or hypercalcemia.

Gamma Interferon

Gamma interferon is also a multifunctional cytokine which has biological activities similar to those of LT and TNF including cytotoxicity in certain tumor cell lines. Nedwin et al. [21] found that gamma interferon enhanced the production of both LT and TNF. However, gamma interferon has opposite effects to those of LT, TNF and the IL-1 molecules on osteoclastic bone resorption [5, 22]. It is an antagonist of bone resorption, and we find it is less effective in inhibiting bone resorption stimulated by parathyroid hormone or 1,25-dihydroxyvitamin D than it is in inhibiting bone resorption stimulated by the cytokines IL-1, LT and TNF [22]. It inhibits bone resorption in both fetal rat long bone organ cultures and neonatal mouse calvarial organ cultures. It appears to mediate its effects by inhibiting differentiation of committed osteoclast progenitors into mature osteoclasts [23]. It also inhibits the proliferation of osteoclast precursors, but much larger concentrations are required. Gamma interferon also has effects on bone collagen synthesis [11]. However, in this situation, it mimics the effects of the cytokines LT, TNF and IL-1 by inhibiting bone collagen synthesis.

The precise role of gamma interferon in normal bone remodeling or in pathologic bone resorption is unknown. Clearly, it may be one member of the network of leukocyte products which affect bone cells. It may have a potential therapy for bone resorption which is mediated by other cytokines such as LT for example in myeloma.

Colony-Stimulating Factors (CSFs)

The weight of evidence suggests that osteoclasts are derived from hematopoietic stem cells and probably from precursors in the marrow which have the potential to develop into cells in the monocyte-macrophage lineage or cells in the granulocyte lineage. Certainly, a number of observations suggest that osteoclast precursors respond to the CSFs. These observations include:

(1) In long-term human marrow cultures, a technique which has been developed by Roodman and co-workers [18, 23], multinucleated cells with osteoclast characteristics form following prolonged culture. This technique is a modified form of the marrow culture system developed by Dexter in which marrow cells are cultured in the presence of an adherent cell layer containing various other cell types including macrophages, fibroblasts and stromal cells. Using this technique, it has been found that recombinant human CSF-GM and purified CSF-1, both CSFs which enhance the proliferation of cells in the monocyte-macrophage lineage, lead to increased formation of cells with osteoclast characteristics [24]. This response occurs after 1 week of incubation with the colony-stimulating activity and is potentiated when 1,25-dihydroxyvitamin D is added during the later periods of the culture. Studies using autoradiography indicate that both CSF-GM and CSF-1 stimulate the formation of osteoclasts by causing increased proliferation of the progenitors, rather than differentiation of the committed precursors.

(2) In a culture system devised by Burger et al. [25] to study the formation of osteoclasts in vitro, sources of colony-stimulating activity derived from fibroblast-conditioned media enhanced the formation of cells with osteoclast characteristics. In this system, exogenous mononuclear cells or periosteal cells are added to organ cultures of 17-day mouse metatarsal bones. Osteoclasts form within the mineralized core of the bones and resorb in the mineralized matrix. However, purified CSF-1 appears to decrease the formation of tartrate-resistant acid phosphatase cells, presumably containing osteoclasts [26].

(3) Studies in a murine variant of osteopetrosis, the op/op mouse, show that colony-stimulating activity may be important in the regulation of bone resorption [27]. In this variant of osteopetrosis, there is a failure of differentiation of cells of the monocyte-macrophage lineage, presumably caused by decreased production of colony-stimulating activity by stromal marrow fibroblasts. When hematopoietic stem cells are cultured from these animals, they respond normally to exogenous colony-stimulating activity in vitro. However, the stromal cells from the affected animals do not produce normal amounts of colony-stimulating activity. When considered together, the results suggest that CSF production in these animals is deficient and is responsible for the impaired osteoclast formation.

(4) There have been a number of descriptions of leukocytosis occurring in both patients and animals with the syndrome of humoral hypercalcemia of malignancy [28–30]. In some of these examples, the tumors have been shown to release a factor with colony-stimulating activity in vitro [31–33]. Nude mice carrying these human tumors also develop leukocytosis. It is possible that the bone-resorbing activity and the colony-stimulating activity may be mediated by the same factor, although colony-stimulating factors do not appear to stimulate osteoclastic bone resorption in vitro. It is more likely that the effects of the CSF and the tumor product enhance each other and are synergistic on osteoclast progenitors and precursors.

1,25-Dihydroxyvitamin D

1,25-Dihydroxyvitamin D should be considered among the factors which regulate normal osteoclast differentiation and formation as well as modulating the production of various cytokines which may be important in bone resorption. 1,25-Dihydroxyvitamin D is a potent stimulator of osteoclastic bone resorption both in vitro and in vivo [34, for review, cf. 35]. There is now well-documented evidence that 1,25-dihydroxyvitamin D enhances the differentiation of cells in the monocyte-macrophage lineage [18, 36] and acts as a differentiation agent for human leukemia cell lines such as HL-60 [37, 38] and U937 cells [39]. Recent information suggests that active metabolites of vitamin D may also influence the differentiation of epidermal cells [40] and active vitamin D metabolites have been associated with improvement in psoriatic lesions [41]. 1,25-Dihydroxyvitamin D also acts as a fusogen for committed osteoclast progenitors. Using the modified Dexter marrow culture system, 1,25-dihydroxyvitamin D in low concentrations enhances the fusion of marrow mononuclear cells to form cells with osteoclast characteristics [18, 42]. 1,25-Dihydroxyvitamin D has little effect on the proliferation of osteoclast progenitors [18, 24]. In support of the potential role for 1,25-dihydroxyvitamin D as an agent which causes differentiation of committed osteoclast progenitors, Key et al. [43] found that in an infant with malignant osteopetrosis, a disorder which is characterized by failure to form competent osteoclasts, that treatment for 3 months with 1,25-dihydroxyvitamin D led to active osteoclasts and bone resorption.

1,25-Dihydroxyvitamin D also has additional effects as an immunoregulatory molecule which could influence bone resorption [for review, cf. 44].In this way, 1,25-dihydroxyvitamin D may influence local regulation of osteoclast function. 1,25-Dihydroxyvitamin D inhibits interleukin-2 production by normal activated lymphocytes, and this leads to impaired lymphocyte mitogenesis [45]. This inhibitory effect of 1,25-dihydroxyvitamin D on lymphocyte mitogenesis can be reversed, at least partially, by adding interleukin-2 back to the cultures [46]. Moreover, 1,25-dihydroxyvitamin D can augment IL-1 production by indirect effects from the leukemic cell line U937 in response to other factors such as T-cell products by causing maturation of these cells [47]. Provvedini et al. [48] have shown receptors for 1,25-dihydroxyvitamin D in T lymphocytes following activation by mitogens such as phytohemagglutinin, and in addition have found 1,25-dihydroxyvitamin D receptors in a number of lymphoid cell lines and cells of the monocyte-macrophage family [44]. Thus, 1,25-dihydroxyvitamin D has multiple effects on immune cell function as well as directly on osteoclastic resorption. It is difficult at the present time to place these within the context of physiological bone remodeling. The recent finding that normal lymphoid cells transformed by inoculation with the HTLV type I virus (human lymphotrophic virus type I) [49] suggests that normal activated lymphocytes in the bone microenvironment also have the potential to produce this mediator as an important regulator of local osteoclast activity.

1,25-Dihydroxyvitamin D also has important and complex effects on cells with the osteoblast phenotype. 1,25-Dihydroxyvitamin D stimulates osteoblast-like cells to secrete the bone Gla protein (osteocalcin) [50]. 1,25-Dihydroxyvitamin D may also increase alkaline phosphatase content in cells with the osteoblast phenotype [51, 52]. Other studies have suggested that 1,25-dihydroxyvitamin D may alter the responsiveness of cells with the osteoblast phenotype to parathyroid hormone, as assessed by their cyclic AMP or adenylate cyclase response [53, 54]. When considered together, these findings suggest that 1,25-dihydroxyvitamin D has an important regulatory influence on maturation of cells in the osteoblast lineage. The effects of 1,25-dihydroxyvitamin D on osteoblasts are important in vivo because deficiency of 1,25-dihydroxyvitamin D leads to impaired mineralization of bone (rickets).

The major site of the production of 1,25-dihydroxyvitamin D is in the proximal tubule of the kidney. However, there is accumulating evidence that 1,25-dihydroxyvitamin D can also be produced in extrarenal sites. 1-Hydroxylase activity has been detected in anephric individuals [55, 56]. 1-Hydroxylase activity has been demonstrated in vitro in tissues including the placenta [57, 58], chick chorion [59], rabbit and human bone cells [60. 61], melanoma cells [62], human pulmonary alveolar macrophages derived from patients with sarcoidosis [63] and sarcoid granulomas [64]. The demonstration that HTLV-transformed lymphocytes can metabolize 25-hydroxyvitamin D to 1,25-dihydroxyvitamin D may explain the mechanism for increased serum 1,25-dihydroxyvitamin D levels in some patients with adult T-cell lymphoma caused by this virus [49] and raises the possibility that normal activated lymphocytes under appropriate circumstances can produce this metabolite.

Transforming Growth Factor Beta

Transforming growth factor (TGF) beta may have important effects in normal bone remodeling. It has powerful effects on both osteoclasts and osteoblasts. In fetal rat long bones, it inhibits osteoclastic bone resorption [65]. This effect appears due to a hydroxyurea-like effect on osteoclast progenitors to inhibit their proliferation. Its effects are most prominent in the later phases of resorption, and this factor has a delayed onset of action as an inhibitor. We have tested its effects on both isolated osteoclasts as well as on the formation of osteoclasts. It inhibits proliferation of osteoclast progenitors and differentiation into mature cells [66]. However, in addition to these effects it also acts on the mature preformed osteoclast to decrease its activity [67]. Although in fetal rat long bones and human systems it acts as an inhibitor of osteoclast activity, in mouse bone cultures it stimulates prostaglandin synthesis and these prostaglandins in turn can lead to bone resorption [68].

TGF beta is enriched in the bone matrix. Seyedin et al. [69] purified TGF beta-1 from demineralized bone matrix and found that it was abundant in this site. This result was confirmed by Hauschka et al. [70]. More recently, Seyedin et al.[71] have purified a second form of TGF beta from the bone matrix. This was initially called cartilage-inducing factor B and is now known as TGF beta-II. TGF beta-II seems to have identical effects to TGF beta-1 on bone culture systems in vitro, although it may be slightly more potent in most assays.

TGF beta is released from bones when they are stimulated to resorb [72]. When bone organ cultures are activated by agents such as PTH, IL-1 or 1,25-dihydroxyvitamin D, the conditioned media bathing the organ cultures contains TGF beta activity. Production of this activity is inhibited when calcitonin, which inhibits osteoclastic bone resorption, is added to the cultures. However, the TGF beta which is released from resorbing bones is in a latent form and requires release from a binding protein for activation [73]. This binding protein is of large molecular weight. TGF beta can be cleaved from its binding protein by the presence of acid, and thus the environment of the ruffled border under the osteoclast may be optimal for causing activation of TGF beta. However, recent evidence also suggests that TGF beta can be activated by osteoclasts independent of acid production [74].

TGF beta also has powerful but confusing effects on cells of the osteoblast phenotype. In fetal rat calvarial cultures, TGF beta in low concentrations causes an increase in both DNA synthesis and in collagen synthesis [73, 75]. However, the increase in collagen synthesis is paralleled by an increase in noncollagen synthesis. The changes in collagen synthesis are more prominent in the periosteal cells but not in cells in the central bone which are osteoblast enriched. On cultured cells with the osteoblast phenotype, TGF beta may inhibit proliferation and stimulate differentiation. At the same time it causes a marked change in the cell shape of these cells associated with an increase in the synthesis of collagen and alkaline phosphatase [73, 75]. However, in other cultured bone cells it may cause the opposite effects, namely an increase in cell proliferation and a decrease in differentiated function.

Platelet-Derived Growth Factor

Platelet-derived growth factor (PDGF) is a circulating mitogen for cells of mesenchymal origin. It has many properties in common with TGF beta. Like TGF beta, it is sequestered within the alpha granules of the platelet and is released when platelets are activated by contact with thrombin, collagen surfaces or subendothelial basement membranes at sites where blood vessels are injured [for general review, cf. 76, 77]. A protein with PDGF-like activity has been described in cells of the monocyte-macrophage family [78, 79]. PDGF has other biological functions in addition to its mitogenic effects. It has chemotactic activity for monocytes, neutrophils and fibroblasts and has effects on bone cells (see below). Analysis of its N-terminal amino acid sequence has demonstrated that it consists of two polypeptide chains linked by disulfide bonds which are encoded by two separate genes. One of these chains, the PDGF-B polypeptide chain, shows extensive sequence homology with P28 sis, the oncogene product of the simian sarcoma virus, an acute transforming primate retrovirus [80, 81]. The effects of PDGF on mesenchymal cells are different from most other growth factors such as EGF, TGF alpha and the somatomedins. PDGF is a competence growth factor, which means that after short exposure to PDGF for 30 min or less, cells become competent to replicate their DNA and divide when PDGF is then removed from the medium. In contrast, growth factors such as EGF, TGF alpha and the somatomedins must be present with cells for more prolonged intervals to sustain a mitogenic response.

DNA probes to the v-sis gene show that this oncogene is frequently expressed in human osteosarcoma cells. Many of these osteosarcoma cells have the osteoblast phenotype. Heldin et al. [82] showed that one human osteosarcoma cell line (I-2 OS) secreted a polypeptide which has similar chemical characteristics, biological activities and chromatographic behavior to human PDGF. Moreover, when this polypeptide was labeled it could be immunoprecipitated by antisera to PDGF, and competed with labeled TGF for binding to cellular receptors. The biological activity of this factor can also be blocked with antisera to PDGF [83]. A polypeptide with PDGF-like activity is enriched in demineralized bovine bone matrix [70]. This peptide is present in bone in levels of 50 ng/g dry bone, There is other evidence that PDGF-like peptides may be important in bone cell function. Tashjian et al. [84] demonstrated that platelet-derived purified PDGF stimulated prostaglandin synthesis in cultured mouse calvaria which led to bone resorption. Canalis [85] showed that purified platelet-derived PDGF could stimulate bone collagen synthesis in rat calvarial organ cultures in vitro. In osteosarcoma cells with the osteoblast phenotype, Graves et al. [86, 87] showed that platelet-derived PDGF has a mitogenic effect. Recently, Valentin-Opran et al. [88] in a preliminary report have suggested that a PDGF-like peptide is produced by normal human bone cells dispersed from trabecular bone surfaces.

In summary, immune cells produce a network of factors which may regulate bone cell activity. These factors may be important in the highly integrated control of normal bone remodelling, and their excess production is now clearly linked to the bone destruction which occurs in certain disease states.

Acknowledgments

Some of the data reviewed here were gathered with the support of grants AR28149, CA40035 and RR01346. We are grateful to Nancy Garrett for her secretarial support.

References

1 Horton JE, Raisz LG, Simmons HA, et al: Bone resorbing activity in supernatant fluid from cultured human peripheral blood leukocytes. Science 1972;177:793–795.
2 Trummel CL, Mundy GR, Raisz LG: Release of osteoclast activating factor by normal human peripheral blood leukocytes. J Lab Clin Med 1975;85:1001–1007.
3 Mundy GR, Raisz LG, Cooper RA, et al: Evidence for the secretion of an osteoclast stimulating factor in myeloma. New Engl J Med 1974;291:1041–1046.
4 Gowen M, Wood OD, Ihrie EJ, et al: An interleukin-1-like factor stimulates bone resorption in vitro. Nature 1983;306:378–380.
5 Gowen M, Mundy GR: Actions of recombinant interleukin-1, interleukin-2 and interferon gamma on bone resorption in vitro. J Immunol 1986;136:2478–2482.
6 Heath JK, Saklatvala J, Meikle MC, et al: Pig interleukin-1 (catabolin) is a potent stimulator of bone resorption in vitro. Calcif Tissue Int 1985;37:95–97.
7 Thomson BM, Saklatvala J, Chambers TJ: Osteoblasts mediate interleukin-1 stimulation of bone resorption by rat osteoclasts. J Exp Med 1986;164:104–112.
8 Sabatini M, Boyce B, Aufdemorte T, et al: Infusions of recombinant human interleukin-1 alpha and beta cause hypercalcemia in normal mice. Proc Natl Acad Sci USA 1988;85:5235–5239.
9 Boyce BF, Aufdemorte T, Garrett IR, et al: Stimulation of bone turnover in vivo by interleukin-1. J Bone Miner Res 1988;3(suppl 1):542.
10 Canalis E: Interleukin-1 has independent effects on deoxyribonucleic acid and collagen synthesis in cultures of rat calvariae. Endocrinology 1986;118:74–81.
11 Smith D, Gowen M, Mundy GR: Effects of interferon gamma and other cytokines on collagen synthesis in fetal rat bone cultures. Endocrinology 1987;120:2494–2499.
12 Pacifici R, Rifas L, Teitelbaum S, et al: Spontaneous release of interleukin-1 from human blood monocytes reflects bone formation in idiopathic osteoporosis. Proc Natl Acad Sci USA 1987;84:4616–4620.
13 Sato K, Fujii Y, Kasono K, et al: Interleukin-1 alpha and PTH-like factor are responsible for humoral hypercalcemia associated with esophageal carcinoma cells (EC-GI). J Bone Miner Res 1987;2 (suppl 1);387.
14 Sammon PJ, Wronski TJ, Flueck JA, et al: Humoral hypercalcemia of malignancy: Evidence for interleukin-1 activity as a bone resorbing factor released by human transitional-cell carcinoma cells; in Cohn (ed): Calcium Regulation and Bone Metabolism. New York, Elsevier, 1987, pp 383–396.
15 Bertolini DR, Nedwin GE, Bringman TS, et al: Stimulation of bone resorption and inhibition of bone formation in vitro by human tumour necrosis factors. Nature 1986;319:516–518.
16 Tashjian AH, Voelkel EF, Lazzaro M, et al: Tumor necrosis factor alpha (cachetin) stimulates bone resorption in mouse calvaria via a prostaglandin-mediated mechanism. Endocrinology 1987;120:2029–2036.
17 Thomson BM, Mundy GR, Chambers TJ: Tumor necrosis factors alpha and beta induce osteoblastic cells to stimulate osteoclastic bone resorption. J Immunol 1987;138:775–779.
18 Roodman GD, Takahashi N, Bird A, et al: Tumor necrosis factor alpha (TNF) stimulates formation of osteoclast-like cell (OCL) in long-term human marrow cultures by stimulating production of interleukin-1 (IL-1). Clin Res 1987;35:515A.

19 Garrett IR, Durie BGM, Nedwin GE, et al: Production of the bone resorbing cytokine lymphotoxin by cultured human myeloma cells. N Engl J Med 1987;317: 526–532.
20 Josse RG, Murray TM, Mundy GR, et al: Observations on the mechanism of bone resorption induced by multiple myeloma marrow culture fluids and partially purified osteoclast activating factor. J Clin Invest 1981;67:1472–1481.
21 Nedwin GE, Svedersky LP, Bringman TS, et al: Effect of interleukin-2, interferon-gamma, and mitogens on the production of tumor necrosis factors alpha and beta. J Immunol 1985;135:2492–2497.
22 Gowen M, Nedwin G, Mundy GR: Preferential inhibition of cytokine stimulated bone resorption by recombinant interferon gamma. J Bone Miner Res 1986;1:469–474.
23 Takahashi N, MacDonald BR, Hon J, et al: Recombinant human transforming growth factor alpha stimulates the formation of osteoclast-like cells in long-term human marrow cultures. J Clin Invest 1986;78:894–898.
24 MacDonald BR, Mundy GR, Clark S, et al: Effects of human recombinant CSF-GM and highly purified CSF-1 on the formation of multinucleated cells with osteoclast characteristics in long-term bone marrow cultures. J Bone Miner Res 1986;1:227–233.
25 Burger EH, Van der Meer JWM, Van de Gevel JS, et al: In vitro formation of osteoclasts from long-term bone cultures of bone marrow mononuclear phagocytes. J Exp Med 1982;156:1604–1614.
26 Uitewaal PHM, Lips P, Netelenbos JC: An analysis of bone structure in patients with hip fracture. Bone and Mineral 1987;3:63–73.
27 Wiktor-Jedrzejzcak W, Ahmed A, Szczylik C, et al: Hematological characterisation of congenital osteopetrosis in op/op mouse. J Exp Med 1982;156:1516–1527.
28 Lee MY, Baylink DJ: Hypercalcemia, excessive bone resorption, and neutrophilia in mice bearing a mammary carcinoma. Proc Soc Exp Biol Med 1983;172:424–429.
29 Oshawa N, Ueyama Y, Morita K, et al: Heterotransplantation of human functioning tumors to nude mice; in Nomura (ed): Proceedings of the Second International Workshop on Nude Mice. Tokyo, University of Tokyo Press, 1977, pp 395–405.
30 Kondo Y, Sato K, Ohkawa H, et al: Association of hypercalcemia with tumors producing colony-stimulating factor(s). Cancer Res 1983;43:2368–2374.
31 Asano S, Urabe A, Okabe T, et al: Demonstration of granulopoietic factor(s) in the plasma of nude mice transplanted with a human lung cancer and in the tumor tissue. Blood 1977;49:845–852.
32 Saito K, Kuratomi Y, Yamamoto K, et al: Primary squamous cell carcinoma of the thyroid associated with marked leukocytosis and hypercalcemia. Cancer 1981; 48:2080–2083.
33 Sato K, Mimura H, Han DC, et al: Production of bone-resorbing activity and colony-stimulating activity in vivo and in vitro by a human squamous cell carcinoma associated with hypercalcemia and leukocytosis. J Clin Invest 1986;78:145–154.
34 Raisz LG, Trummel CL, Holick MF, et al: 1,25-Dihydroxycholecalciferol: a potent stimulator of bone resorption in tissue culture. Science 1972;175:768–769.
35 Mundy GR, Roodman GD: Osteoclast ontogeny and function: in Peck (ed): Bone and Mineral Research. New York, Elsevier, 1987, vol V, pp 209–280.

36 Abe E, Miyaura C, Sakagami H, et al: Differentiation of mouse myeloid leukemia cells induced by 1α,25-dihydroxyvitamin D_3. Proc Natl Acad Sci USA 1981;78:4990–4994.
37 Miyaura C, Abe E, Nomura H, et al: 1α,25-Dihydroxyvitamin D_3 suppresses proliferation of murine granulocyte-macrophage progenitor cells (CFU-C). Biochem Biophys Res Commun 1982;108:1728–1733.
38 Reitsma PH, Rothberg PG, Astria SM, et al: Regulation of myc gene expression in HL-60 leukaemia cells by a vitamin D metabolite. Nature 1983;306:492–494.
39 Dodd RC, Cohen MS, Newman SL, et al: Vitamin D metabolites change the phenotype of monoblastic U937 cells. Proc Natl Acad Sci USA 1983;80:7538–7541.
40 Hosomi J, Hosoi J, Abe E, et al: Regulation of terminal differentiation of cultured mouse epidermal cells by 1α,25-dihydroxyvitamin D_3. Endocrinology 1983;3:1950–1957.
41 Morimoto S, Onishi T, Imanaka S, et al: Topical administration of 1,25-dihydroxyvitamin D_3 for psoriasis: Report of five cases. Calcif Tissue Int 1986;38:119–122.
42 Ibbotson KJ, Roodman GD, McManus LM, et al: Identification and characterization of osteoclast-like cells and their progenitors in cultures of feline marrow mononuclear cells. J Cell Biol 1984;94:471–480.
43 Key L, Carnes D, Cole S, et al: Treatment of congenital osteopetrosis with high dose calcitriol. N Engl J Med 1984;310:410–415.
44 Manolagas SC, Provvedini DM, Tsoukas C: Interactions of 1,25-dihydroxyvitamin D_3 and the immune system. Mol Cell Endocrinol 1985;43:113–122.
45 Tsoukas CD, Provvedini DM, Manolagas SC: 1,25-Dihydroxyvitamin D_3: A novel immunoregulatory hormone. Science 1984;224:1438–1440.
46 Rigby WFC, Stacy T, Fanger MW: Inhibition of T lymphocyte mitogenesis by 1,25-dihydroxyvitamin D_3 (calcitriol). J Clin Invest 1984;74:1451–1455.
47 Amento EP, Bhalla AK, Kurnick JT, et al: 1α,25-Dihydroxyvitamin D_3 induces maturation of the human monocyte cell line U937 and, in association with a factor from human T lymphocytes, augments production of the monokine, mononuclear cell factor. J Clin Invest 1984;73:731–739.
48 Provvedini DM, Tsoukas CD, Deftos LJ, et al: 1,25-Dihydroxyvitamin D_3 receptors in human leukocytes. Science 1983;221:1181–1183.
49 Fetchick DA, Bertolini DR, Sarin P, et al: Production of 25-hydroxyvitamin D by human T cell lymphotrophic virus-transformed cord blood lymphocytes. J Clin Invest 1986;78:592–596.
50 Price PA, Baukol SA: 1,25-Dihydroxyvitamin D_3 increases synthesis of the vitamin K-dependent bone protein by osteosarcoma cells. J Biol Chem 1980;255:11660–11663.
51 Manolagas SC, Burton DW, Deftos LJ: 1,25-Dihydroxyvitamin D_3 stimulates the alkaline phosphatase activity of osteoblast-like cells. J Biol Chem 1981;256:7115–7117.
52 Rodan GA, Rodan SB: Expression of the osteoblastic phenotype; in Peck (ed): Bone and Mineral Research; Annual 2. New York, Elsevier, 1984, pp 244–285.
53 Catherwood BD: 1,25-Dihydroxycholecalciferol and glucocorticosteroid regulation of adenylate cyclase in an osteoblast-like cell line. J Biol Chem 1985;260:736–743.
54 Kubota M, Ng KW, Martin TJ: Effect of 1,25-dihydroxyvitamin D_3 on cyclic AMP responses to hormones in clonal osteogenic sarcoma cells. Biochem J 1985;231:11–17.

55 Lambert PW, Stern PH, Avioli RC, et al: Evidence for extrarenal production of 1α,25-dihydroxyvitamin D in man. J Clin Invest 1982;69:722–725.
56 Barbour GL, Coburn JW, Slatopolsky E, et al: Hypercalcemia in an anephric patient with sarcoidosis: Evidence for extrarenal generation of 1,25-dihydroxyvitamin D. N Engl J Med 1981;305:440–443.
57 Weisman Y, Harell A, Edelstein S, et al: 1α,25-Dihydroxyvitamin D_3 and 24,25-dihydroxyvitamin D_3 in vitro synthesis by human decidua and placenta. Nature 1979;281:317–319.
58 Gray TK, Lester GE, Lorenc RS: Evidence for extrarenal 1α-hydroxylation of 25-hydroxyvitamin D_3 in pregnancy. Science 1979;204:1311–1313.
59 Puzas JE, Turner RT, Forte MD, et al: Metabolism of $25(OH)D_3$ to $1,25(OH)_2D_3$ and $24,25(OH)_2D_3$ by chick chorioallantoic cells in culture. Gen Comp Endocrinol 1980;42:116–122.
60 Turner RT, Puzas JE, Forte MD, et al: In vitro synthesis of 1α,25-dihydroxycholecalciferol and 24,25-dihydroxycholecalciferol by isolated calvarial cells. Proc Natl Acad Sci USA 1980;77:5720–5724.
61 Howard GA, Turner RT, Sherrard DJ, et al: Human bone cells in culture metabolize 25-hydroxyvitamin D_3 to 1,25-dihydroxyvitamin D_3 and 24,25-dihydroxyvitamin D_3. J Biol Chem 1981;256:7738–7740.
62 Frankel TL, Mason RS, Hersey P, et al: The synthesis of vitamin D metabolites by human melanoma cells. J Clin Endocrinol Metabol 1983;57:627–631.
63 Adams JS, Sharma OP, Gacad MA, et al: Metabolism of 25-hydroxyvitamin D_3 by cultured pulmonary alveolar macrophages in sarcoidosis. J Clin Invest 1983; 72:1856–1860.
64 Mason RS, Frankel T, Chan YL, et al: Vitamin D conversion by sarcoid lymph node homogenate. Ann Intern Med 1984;100:59–61.
65 Pfeilschifter J, Bonewald L, Mundy GR: TGFB is released from bone with one or more binding proteins which regulate its activity. J. Bone Miner Res 1987;2(suppl 1):249.
66 Chenu C, Pfeilschifter J, Mundy GR, Roodman GD: Transforming growth factor β inhibits formation of osteoclast-like cells in long-term human marrow cultures. Proc Natl Acad Sci USA 1988;85:5683–5687.
67 Oreffo ROC, Bonewald L, Garrett IR, et al: Transforming growth factors beta-I and II inhibit osteoclast activity. J Bone Miner Res 1988;3(suppl 1):439.
68 Tashjian AH, Voelkel EF, Lloyd W, et al: Actions of growth factors on plasma calcium. J Clin Invest 1986;78:1405–1409.
69 Seyedin SM, Thomas TC, Thompson AY, et al: Purification and characterization of two cartilage-inducing factors from bovine demineralized bone. Proc Natl Acad Sci USA 1985;82:2267–2271.
70 Hauschka PW, Mavrakos AE, Iafrati MD, et al: Growth factors in bone matrix. J Biol Chem 1986;261:12665–12674.
71 Seyedin SM, Thompson AY, Bentz H, et al: Cartilage-inducing factor A. J Biol Chem 1987;13:5693–5695.
72 Pfeilschifter J, Mundy GR: Modulation of transforming growth factor beta activity in bone cultures by osteotropic hormones. Proc Natl Acad Sci USA 1987;84:2024–2028.
73 Pfeilschifter J, D'Souza SM, Mundy GR: Effects of transforming growth factor beta on osteoblastic osteosarcoma cells. Endocrinology 1987;121:212–218.

74 Oreffo ROC, Mundy GR, Bonewald LB: Osteoclasts activate latent transforming growth factor beta and vitamin A treatment increases TGF beta activation. Calcif Tissue Int 1988:42(suppl)56.
75 Noda M, Rodan GA: Type-beta transforming growth factor inhibits proliferation and expression of alkaline phosphatase in murine osteoblast-like cells. Biophys Biochem Res Commun 1986;140:56–65.
76 Cochran BH: The molecular action of platelet-derived growth factor; in Klein (ed): Advances in Cancer Research. New York, Academic Press, 1985, vol 45, pp 183–216.
77 Stiles CD: The molecular biology of platelet-derived growth factor. Cell 1983; 33:653–655.
78 Mornex JF, Martinet Y, Yamauchi K, et al: Spontaneous expression of the c-sis gene and release of a platelet-derived growth factor-like molecule by human alveolar macrophages. J Clin Invest 1986;78:61–66.
79 Shimodako K, Raines EW, Madtes DK, et al: A significant part of macrophage-derived growth factor consists of at least two forms of PDGF. Cell 1985;43:277–286.
80 Deuel TF, Huang JS, Huang SS, et al: Expression of a platelet-derived growth factor-like protein in simian sarcoma virus transformed cells. Science 1983;221:1348–1350.
81 Doolittle RF, Hunkapiller MW, Hood LE, et al: Simian sarcoma virus onc gene, v-sis, is derived from the gene (or genes) encoding a platelet-derived growth factor. Science 1983;221:275–277.
82 Heldin CH, Westermark B, Wasteson A: Specific receptors for platelet-derived growth factor on cells derived from connective tissue and glia. Proc Natl Acad Sci USA 1981;78:3664–3668.
83 Graves DT, Owen AJ, Antoniades HN: Evidence that a human osteosarcoma cell line which secretes a mitogen similar to platelet-derived growth factor requires growth factors present in platelet-poor plasma. Cancer Res 1983;43:83–87.
84 Tashjian AH, Hohmann EL, Antoniades HN, et al: Platelet-derived growth factor stimulates bone resorption via a prostaglandin mediated mechanism. Endocrinology 1982;111:118–124.
85 Canalis E: Platelet-derived growth factor stimulates DNA and protein synthesis in cultured fetal rat calvaria. Metabolism 1980;30:970–975.
86 Graves DT, Owen AJ, Barth RK, et al: Detection of c-sis transcripts and synthesis of PDGF-like proteins by human osteosarcoma cells. Science 1984;226:972–974.
87 Graves DT, Antoniades HN, Williams SR, et al: Evidence for functional platelet-derived growth factor receptors on MG-63 human osteosarcoma cells. Cancer Res 1984;44:2966–2970.
88 Valentin-Opran A, Delagado R, Valente T, et al: Autocrine production of platelet-derived growth factor (PDGF)-like peptides by cultured normal human bone cells. J Bone Miner Res 1987;2(suppl)254.

Dr. Gregory R. Mundy, Department of Medicine, Division of Endocrinology and Metabolism, University of Texas Health Science Center, 7703 Floyd Curl Drive, San Antonio, TX 78284-7877 (USA)

Sorg C (ed): Macrophage-Derived Cell Regulatory Factors.
Cytokines. Basel, Karger, 1989, vol 1, pp 54–73

Macrophage-Induced Angiogenesis: A Review[1]

Peter J. Polverini

Department of Pathology, Northwestern University Medical and Dental Schools, Chicago, Ill., USA

Introduction

Angiogenesis, the process which leads to the formation of new capillary blood vessels or neovascularization, is an essential component of a wide array of physiological and pathological processes [1, 2]. For example, significant capillary growth occurs transiently during ovulation. The developing fetus and its supporting structures are accompanied by extensive neovascularization which persists throughout embryonic development. The fibrovascular proliferative response that characterizes the formation of inflammatory granulation tissue is obligatory for normal tissue repair. The pathogenesis of a number of inflammatory, degenerative and developmental disease processes can be attributed in part to inappropriate or prolonged neovascularization. Such disorders include, among others, retrolental fibroplasia and diabetic retinopathy, rheumatoid arthritis, periodontitis and a variety of autoimmune processes, psoriasis, progressive systemic sclerosis, and certain congenital vascular anomalies. Lastly, it is well established that the development and progression of solid tumors is an angiogenesis-dependent process [1–3].

The mechanisms of neovascularization and the mediators orchestrating this complex process are as diverse as are the settings in which angiogenesis is encountered. Factors which have been reported to induce or modulate angiogenesis include prostaglandins, heparin, copper ions and copper-binding proteins, cyclic nucleotides, lactic acid, and lowered oxygen tension, several well-characterized polypeptide growth factors, proteolytic enzymes, tissue extracts, histamine and other vasoactive amines, interferon, and

[1] Supported in part by National Institutes of Health Research Grant RO1-HL39926.

several low moleclular weight endothelial mitogens and chemotactic factors [4]. Vascular proliferation has also been induced in several experimental models by certain adult and embryonic tissues and a variety of blood leukocytes [5–7]. It has been during the last decade, however, that the macrophage has emerged as an angiogenesis effector cell of major importance.

It is now well established that the functional domain of the macrophage extends far beyond its originally recognized role as a scavenger cell. Its rich array of secretory products, now numbering approximately 100 well-characterized molecules, anatomic diversity and functional heterogeneity is unmatched by any other cell type [8]. As a result of this remarkable versatility, the macrophage is able to influence every facet of the immune response and inflammation as well as playing a central role in the etiology and/or pathogenesis of a number of disease processes.

Our studies of macrophage-induced angiogenesis initially evolved from our interests in the mechanisms of vascular proliferation during chronic immune-mediated inflammatory responses. At the time there were very little data on endothelial proliferation induced by nonneoplastic stimuli. Studies of tumor neovascularization had firmly established a role for mediators of angiogenesis. In contrast, comparable mechanistic data on endothelial proliferation induced by nonneoplastic stimuli were not available. The occurrence of endothelial proliferation during immunologic reactions had been documented in several studies. Light microscopic and ultrastructural studies of delayed-type hypersensitivity reactions had reported the occurrence of activated and dividing endothelial cells [9, 10]. Graham and Shannon [11] reported that endothelial mitoses were observed during active lymphocyte migration in arthritic joints of rabbits. Anderson et al. [12] demonstrated extensive proliferation of postcapillary venular endothelium in lymph nodes draining skin allographs. Sidkey and Auerbach [5] had shown that capillary proliferation occurred during local graft-versus-host reactions in the skin of mice and suggested that this angiogenic response was induced by the immunocompetent donor lymphocytes. Also, in vitro studies of macrophages indicated that these cells produced growth inhibitory and stimulatory factors for several cell types [13–18]. We therefore initiated a series of experiments to examine and quantitate the vasoproliferative response in two models of delayed-type hypersensitivity: the tuberculin reaction and contact sensitivity to dinitrochlorobenzene [17]. Using a quantitative autoradiographic technique we demonstrated that incorporation of tritiated thymidine by microvascular endothelial cells in the sensitized skin of guinea pigs coincided with the onset and magnitude of mononuclear infiltration at the reaction site. The mechanisms responsible for endothelial proliferation were at the time conjectural: two explanations were postulated. First, endothelial replication was a reparative response to nonspecific injury and necrosis induced by humoral or cell-derived cytotoxins. Alternatively, the proliferation was mediated by growth factors produced by one or more cell types comprising the infiltrate.

One of the first studies implicating macrophages in nonlymphoid mesenchymal cell proliferation was the work of Leibovich and Ross [17]. They demonstrate that in skin wounds of guinea pigs made monocytopenic with hydrocortisone acetate and antimacrophage serum there was a marked delay in healing which manifested as prolonged accumulation of cell debris, reduced fibrosis and a decrease in the number of fibroblasts at the wound site. To examine this apparent relationship between macrophages and connective tissue repair more directly, these workers tested the ability of macrophage-conditioned media to stimulate the growth of skin wound fibroblasts in culture [18]. Their results showed that macrophages retrieved from the unstimulated or inflamed peritoneal cavity of guinea pigs produced a potent growth factor for fibroblasts. The potency of the growth-providing effect appeared to be dependent upon the culture duration, and the density of macrophages. In addition, this growth stimulation was augmented with oil-induced macrophages or when cells were challenged with latex in culture. These observations were later widely confirmed in a number of other laboratories [20].

With this background information we asked whether macrophages might have a similar growth-promoting effect on vascular endothelium in vivo. At about the time we began our studies, Clark et al. [21] published a report showing that wound macrophages when introduced into rabbit corneas stimulated neovascularization in the wake of an acute inflammatory response. Using the corneal bioassay of neovascularization we examined whether macrophages or their conditioned culture media, when introduced into guinea pig corneas, could stimulate ingrowth of capillary blood vessels [7]. Peritoneal macrophages obtained from Balb/c mice and Hartley albino guinea pigs by lavage or following injection of an inflammatory stimulant were processed by standard separation technique to yield a preparation consisting of 85–90% macrophages. Intracorneal injection of viable macrophages or their conditioned culture media were potently angiogenic in over 75% of corneas tested. In contrast, unactivated resident macrophages were either weakly angiogenic or showed no activity at all. A similar preparation of macrophages from inbred strain II guinea pig yielded essentially similar results, thus precluding the possibility that neovascularization occurred as a result of an immunologically mediated inflammatory response. The requirement that macrophages be activated for expression of angiogenic activity was demonstrated with cultured cells. Brief exposure of normally nonangiogenic-resistant peritoneal macrophages to latex induced them to express angiogenesis. Similarly aliquots of cell-free, dialyzed and concentrated culture media from in vivo or in vitro activated macrophages also induced angiogenic activity when incorporated into Hydron polymer and implanted intracorneally. These responses exhibited the same pattern of capillary growth as was observed with viable macrophages. Collectively these results indicated that activated macrophages were able to induce angiogenesis in the absence of inflammation through an inducible secreted product. These studies were subsequently confirmed by Thakral et al. [22] and Hunt et al. [23] with wound-derived macrophages and wound fluids rich in macrophages, and by Moore and Sholley [24] with autologous rabbit peritoneal macrophages.

Pathophysiology of Macrophage-Induced Angiogenesis

Several studies have suggested a relationship between macrophages and angiogenesis in the pathogenesis of certain disease processes. A relationship between the macrophage content of tumors, the rate of tumor growth and the extent of their vascularization has been proposed in several studies. For example, Evans [25, 26] has shown that mice depleted of macrophages by whole-body X-irradiation or azathioprine administered before or after implantation of a syngeneic fibrosarcoma showed a delay in the appearance of tumors, and a marked reduction in tumor vascularization. Mostafa et al. [27, 28] and Stenzinger et al. [29] showed that vascularization of several human tumor cell lines grown on the chorioallantoic membrane of the chick embryo or subcutaneously in nude mice occurred coincidentally with mononuclear cell infiltration at the tumor site. These workers speculated that tumor growth might be partially dependent upon the angiogenic activity of infiltrating macrophages.

To more directly test this association between macrophages and tumor neovascularization we isolated macrophages from a transplantable rat fibrosarcoma and examined them and their 72-hour serum-free conditioned media for angiogenic activity in rat corneas. Our results showed that tumor-associated macrophages (TAM) and their conditioned media were potently angiogenic in vivo and stimulated proliferation of bovine aortic endothelial cells in culture [30]. Moreover, when TAM were combined with tumor cells at a concentration equivalent to the number of macrophages originally present in the tumor there was a marked enhancement of tumor neovascularization and growth. We have also observed a significant reduction in thymidine incorporation by buccal pouch carcinoma cells and capillary endothelium in hamsters depleted of monocytes and macrophages with hydrocortisone and antimacrophage serum [31]. These results would therefore suggest that in certain instances TAM may indirectly stimulate tumor growth by augmenting tumor neovascularization.

Excessive production of angiogenic factors, possibly of monocyte/macrophage origin, have been implicated in the pathogenesis of rheumatoid arthritis [32–34]. This systemic disease is characterized principally by proliferation and thickening of the synovial membrane of diarthrodial joints. The resultant synovial pannus subsequently invades and degrades cartilage and bone and results inevitably in joint destruction [35]. Both persistent neovascularization and sustained monocyte/macrophage infiltration are two of the most prominent features of this disease process [36]. Koch et al. [37] have shown that macrophage-enriched subpopulations from dissociated human rheumatoid synovium are potently angiogenic in the rat corneal bioassay of neovascularization. Interestingly, of the three subpopulations of macrophages isolated, only one fraction was found to consistently induce neovascularization. These findings suggested that macrophages present in diseased synovium are functionally heterogeneous and that the proportion of macrophages expressing this angiogenic activity in a particular diseased synovium may dictate the extent of neovascularization in that tissue and possibly the eventual course of the disease.

The foregoing data strongly suggest an important link between macrophages and neovascularization in at least two disease processes. Whether or not a similar relationship exists in other inflammatory and degenerative disorders remains to be shown.

Regulation of Macrophage Angiogenic Activity

Many of the functions displayed by macrophages are not constitutive but rather are acquired in response to discrete, extracellular signals [38]. This process, termed activation, results in the expression of new or enhanced functional properties. Some well-known activation-dependent functions include presentation of antigen to certain classes of T lymphocytes, the generation and secretion of reactive oxygen intermediates, and destruction of tumor cells and intracellular pathogens. The requirement that macrophage must first be activated to express angiogenic activity was demonstrated early on by several groups of investigators. The environment of healing wounds and inflammatory exudates are normally rich in potent mediators of activation and macrophages obtained from these sources are invariably angiogenic [17, 21–23]. Once activated, cultured macrophages maintain their ability to induce angiogenesis for an extended period of time. Under optimal culture conditions angiogenic activity can be detected in the culture media of activated macrophages, when assayed at 24- to 48-hour intervals, for up to 7 days with little diminution in activity [Polverini, unpubl. observations]. This is in distinction to cytotoxic activity which is only transiently expressed and rapidly decays, within 72 h, when macrophages are cultured under similar conditions.

In addition to activation, the stage of differentiation appears to be an important determinant in the expression of macrophage angiogenic activity. Koch et al. [39] showed that freshly isolated human monocytes and monocyte-derived macrophages are normally unable to induce angiogenesis. However, when these cells were allowed to differentiate in culture and were treated with the activating agents concanavalin A (Con A) or lipopolysaccharide (LPS) they rapidly acquired angiogenic activity. Similarly, we observed that certain cell lines that exhibit properties characteristic of normal macrophages could be induced to express or show enhanced angiogenic activity following exposure to differentiation and/or activation stimuli [40]. The P388D1 and J-774 cells are continuous lines of murine macrophage-like cells which display many properties characteristic of mature activated macrophages while the U-937 and HL60 cells are considered immature human monocyte and myelogenous precursors, respectively, which lack most of the properties of mature macrophages yet can be induced to express a number of macrophage traits when treated with differentiating and activating agents such as LPS or phorbol myristate acetate (PMA). When P388D1, J-774 cells or their 72-hour serum-free conditioned media were introduced intracorneally or incubated with bovine aortic endothelial cells in culture they potently stimulated cornea neovascularization and endothelial cell proliferation. In contrast, when U-937 and HL60 cells or their media were assayed for angiogenic and endothelial cell mitogenic activity they stimulated weak or no detectable angiogenic responses and failed to stimulate proliferation of endothelial cells. However, when these cells were exposed to

PMA or sequentially to PMA and LPS they rapidly differentiated into macrophage-like cells and, in the case of U-937 cells, exhibited enhanced angiogenic and endothelial cell mitogenic activity.

The significance of these in vitro observations to the in vivo setting is not yet clear. There have, however, been several reports implicating two important environmental stimuli in the induction and regulation of macrophage angiogenic activity in vivo. Knighton et al. [41] have reported that tissue oxygen tension exerts a profound influence on macrophage angiogenic activity. They showed that when macrophages were cultured under conditions of hypoxia similar to those present in the center of wounds or a solid tumor, they secrete an active angiogenesis factor in their culture media. When cells were cultured at a PO_2 similar to that of arterial blood (10%) or to that of tissue (5%), macrophages did not induce significant angiogenesis. This suggests that the O_2 gradient that develops in healing wounds may act to regulate tissue vascularization and production of angiogenesis factors by macrophages. A similar autoregulatory mechanism for wound angiogenesis has been proposed for lactate. In addition to low oxygen tension, the wound environment is also typically high in lactate concentration and has a low pH. Jensen et al. [42] showed that concentrated media from rabbit bone marrow derived from macrophages cultured in 15–25 m*M* lactate secreted an angiogenic factor that induced neovascularization in rabbit corneas. When exposed to 1.5 m*M* lactate, macrophages secreted no detectable angiogenic activity. No enhancement of angiogenesis was observed with either structurally similar pyruvate or low pH. These workers suggested that with the return of the normal blood supply in healing wounds, stimuli such as low O_2 tension and lactate would be removed, thereby effecting autoregulation of wound angiogenesis.

Recently there has been considerable interest in the role of transforming growth factor beta (TGFβ) in angiogenesis and wound repair [43]. This polypeptide was first identified by virtue of its ability, in cooperation with transforming growth factor alpha (TGFα)/epidermal growth factor (EGF), to reversibly induce the transformed phenotype in NRK rat fibroblasts. It is now known to have a much broader spectrum of stimulatory and inhibitory effects depending on the type of cells and presence of other cytokines [44, 45]. In vivo TGFβ has been reported to promote wound healing by accelerating collagen production and by inducing fibroplasia and neovascularization [46–48]. In addition, this substance is chemotactic for monocytes and macrophages and may play an important role in regulating production of macrophage growth factors during wound repair [4, 49].

When TGFβ is introduced into rodent corneas, on the chick chorioallantoic membrane or subcutaneously into newborn mice it induced neovascularization coincidentally with prominent mononuclear inflammatory cell infiltration. It has been postulated that TGFβ mediates angiogenesis indirectly by stimulating the release of angiogenic factors from another source, possibly macrophages. Platelets which are a rich source of TGFβ release this mediator from their α-granules during degranulation [50]. The resultant chemotactic gradient that might develop at sites of injury could efficiently attract monocytes and macrophages and allow these cells to express their angiogenic potential. An analogous situation might exist during solid tumor formation where tumor-derived TGFβ could recruit monocytes and macrophages and stimulate production of angiogenic factors [51, 52].

We have found [D.M. Wiseman et al., manuscript submitted] that in addition to being chemotactic for monocytes and macrophages, TGFβ potently activates macrophages for expression of angiogenic activity. Exposure of serum-free cultures of human monocyte-derived macrophages or unstimulated Balb/c mouse peritoneal macrophages for 16–24 h to TGFβ at concentrations as low as 0.64 p*M* induces them to express angiogenic activity in vivo (fig. 1) and to stimulate monolayers of bovine adrenal gland capillary endothelial cells grown on collagen gels to form tube-like structures in vitro (fig. 2a, b).

Another interesting and paradoxical feature of TGFβ is that in vitro it has been shown to inhibit endothelial cell proliferation [53, 54]. Since TGFβ is a bifunctional peptide, its in vitro effects are dependent on a number of factors including the presence of other growth factors and whether or not cells are attached to a substratum. Folkman and Klagsbrun [4] speculate that the growth inhibitory effect of TGFβ on endothelial cells may be an effective mechanism for limiting the action of angiogenic factors and thus regulate the extent of neovascularization.

While there have been numerous detailed studies of the relationship of differentiation and activation to the acquisition of macrophage angiogenic activity and the role of macrophage-derived mediators of neovascularization, the genetic and molecular mechanisms which regulate expression of these vital functions are still poorly understood. With the exception of transformed and neoplastic cells, expression of angiogenic activity is normally held tightly in check. Only when cells are appropriately stimulated, as in the case of activated, differentiated macrophages, do they produce angiogenic factors. We have recently begun to investigate the nature of these genetic controls using the technique of somatic cell hybridization. The

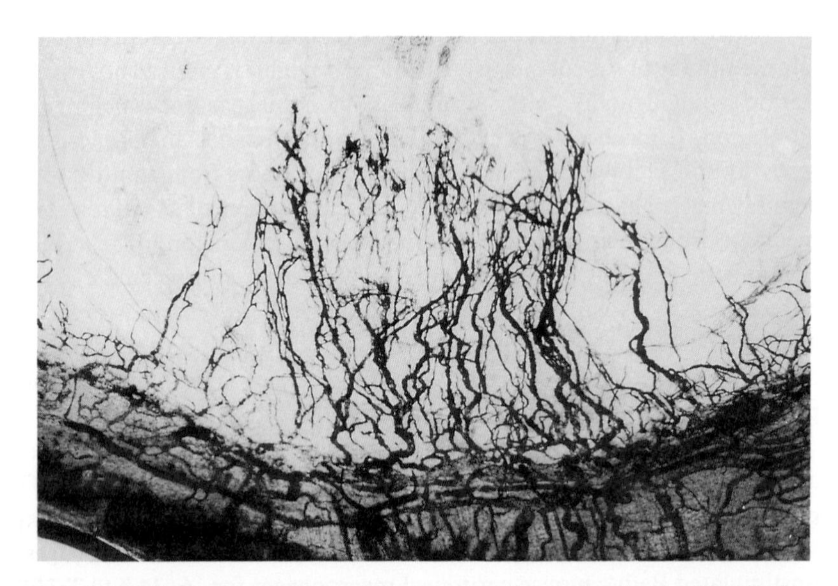

Fig. 1. Colloidal carbon perfused F344 rat cornea 7 days after implanting a Hydron pellet containing 20× concentrated 48-hour serum-free conditioned media from cultures of Balb/c peritoneal Mϕ after a 24-hour exposure to 1 pg of TGFβ. Note the dense brush-like network of capillary sprouts extending from the limbus into the corneal stroma. ×33.

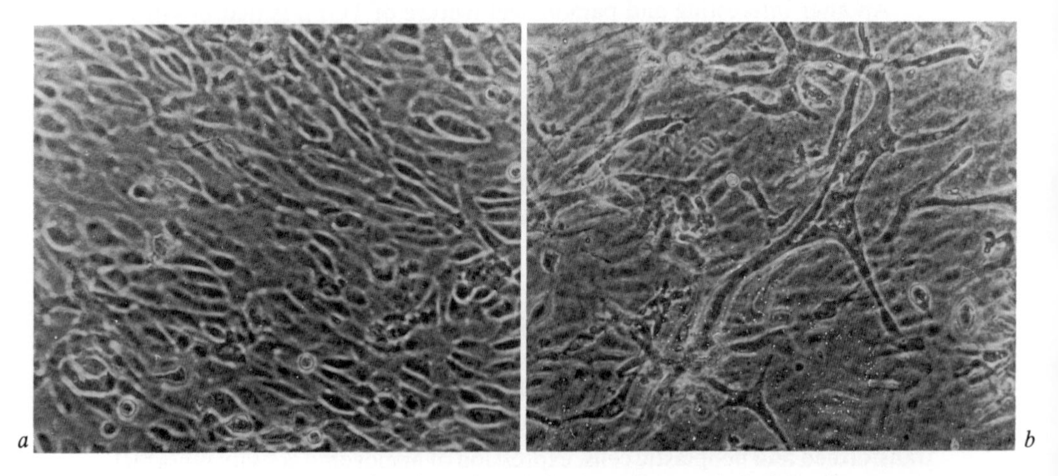

Fig. 2. a Monolayer of bovine adrenal gland capillary endothelial cell grown on dermal collagen and incubated for 24 h with Balb/c mouse unstimulated peritoneal Mϕ-conditioned medium. ×250. *b* Induction of capillary tube-like structures following exposure of the endothelial monolayer to TGFβ-activated Mϕ-conditioned medium. ×250.

hypothesis underlying this work is that the ability of activated macrophages to express angiogenic activity is a recessive trait. A precedent for such a mechanism has recently been documented by Bouck et al. [55] for transformed cell-mediated angiogenesis. In this study we showed that in hybrid cells derived from fusions between angiogenic positive (Ang$^+$) chemically transformed hamster fibroblasts and angiogenic negative (Ang$^-$) normal human fibroblasts, this trait was suppressed. Suppression was shown to be associated with retention of normal human chromosome 1 in these hybrids. This indicated that angiogenic activity is a recessive trait in normal human fibroblasts. We have examined expression of angiogenic activity in several hybrid clones derived from fusions between Ang$^+$ activated mouse peritoneal and mature macrophage cell lines and Ang$^-$ unstimulated peritoneal macrophages or immature macrophage precursor cells and find that angiogenic activity is suppressed in these hybrid cells (table 1, fig. 3a, b). The mechanism underlying suppression of this trait in these hybrids is currently under investigation. This approach may provide an opportunity to analyze the genetic and molecular controls that regulate this complex function in normal cells and perhaps will provide insights into the mechanism underlying the apparent deregulation of angiogenic activity and neoplastic cells.

Macrophage-Derived Mediators of Angiogenesis

The observation that angiogenesis can be induced by macrophage culture fluids suggests that a diffusible angiogenesis factor(s) is produced by these cells. The work of Leibovich and Ross [18, 19] had indicated earlier that activated macrophages produced and secreted growth factor for fibroblasts. This biological activity, termed macrophage-derived growth factor (MDGF), was subsequently confirmed by a number of other laboratories with macrophages from diverse sources and shown to consist of a group of closely related if not identical molecules [20]. In addition to stimulating fibroblast growth, these MDGFs were shown to stimulate the growth of a variety of other nonlymphoid mesenchymal cells including vascular endothelium. In some cases, MDGFs were reported to be angiogenic in vivo [20]. The majority of MDGFs are anionic polypeptides with a molecular weight range from 14 to 80 kd; although it has recently been reported that a component of MDGF may be related to cationic platelet-derived growth factor [56]. They are relatively heat stable, trypsin sensitive and resistant to DNase, RNase and serine esterase inhibitors. They have at least one disulfide bond which is

Table 1. Expression of angiogenic activity by macrophage (M/o) hybrids

Conditioned media from hybrid clones	Proportion of positive responses	
	n	%
Ang^{+} Mϕ × Ang^{-} Mϕ		
$P388D1^{HPRT^{-}Neor}$ × HL60[1]		
F1–1	0/3	0
F1–2	0/4	0
F1–4	2/4	50[2]
$P388D1^{HPRT^{-}}$ × Balb/c UPMϕ		
F–2	0/4	0
F2–5	0/6	0
F2–7	0/7	0
Balb/c LPS APMϕ × $HL60^{HPRT^{-}}$		
F3–2	0/5	0
F3–4	0/3	0
Controls		
$P388D1^{HPRT^{-}}$ × $P388D1^{Neor}$		
F1–9	6/6	100
$HL60^{HPRT^{-}}$ × Balb/c UPMϕ		
F3–7	0/4	0

[1] P388D1 cells resistant to 6-thioguanine ($HPRT^{-}$) and/or G418 (pSV2Neo) were fused to HL60 cells and Balb/c unactivated peritoneal (UP) Mϕ, and $HL60^{HPRT^{-}}$ cells were fused to LPS activated peritoneal (AP) Mϕ using polyethylene glycol (PEG) 1000. Hybrids were selected in HAT with or without G418, clones were isolated, grown to confluence and 48 h serum-free conditioned medium was prepared. Media was concentrated 20-fold by ultrafiltration, incorporated into Hydron and implanted ıntracorneally.
[2] Expression of angiogenic activity was associated with substantial reduction in the chromosome content of these hybrids.

required for biological activity since mitogenic activity is destroyed by treatment with mercaptoethanol and dithiothreitol. The MDGFs also appear to be biochemically and functionally distinct from interleukin-1 (IL-1). In the absence of further information regarding their biochemical identity, the relationship of a number of these MDGFs to macrophage-angiogenic factor(s) and their in vivo mechanism of action remains speculative.

Another group of angiogenic factors recently implicated in macrophage-induced angiogenesis are the fibroblast growth factors (FGFs). These are a family of closely related heparin-binding polypeptide mito-

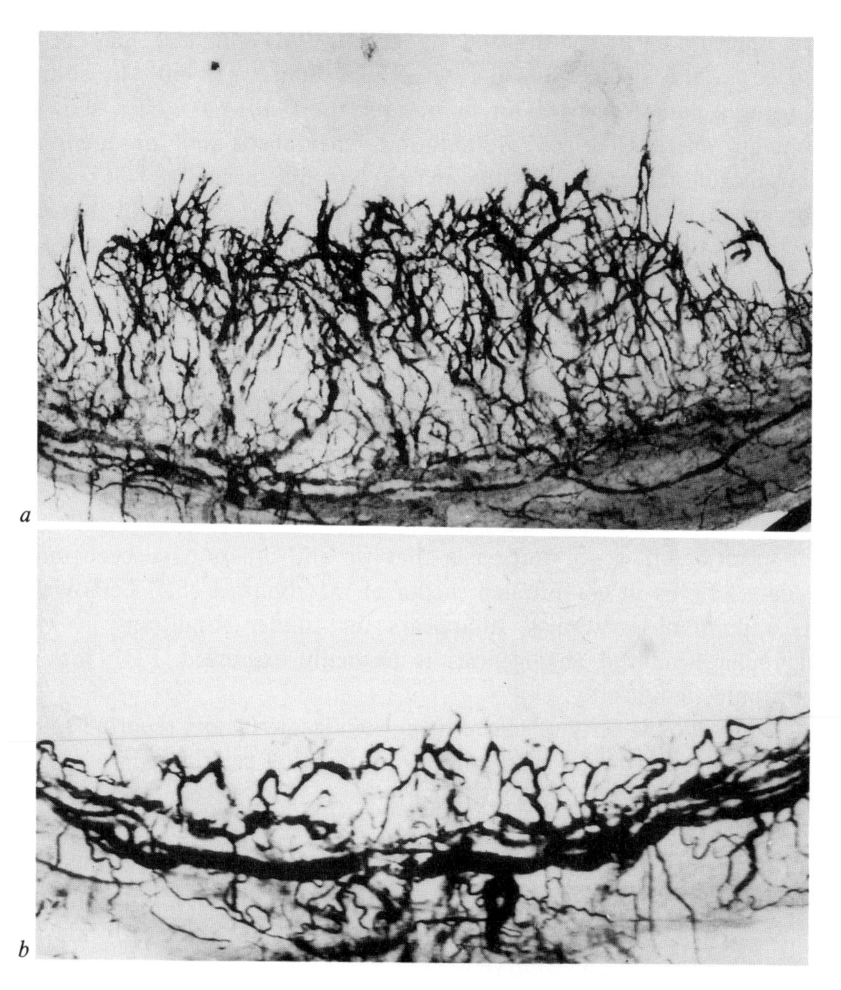

Fig. 3. a, b Colloidal carbon perfused corneas showing positive *(a)* and negative *(b)* neovascular responses 7 days after implanting a Hydron pellet containing 20 × concentrated serum-free conditioned media from P388D1$^{HPRT^-}$ cells *(a)* and hybrid clone F2-5 (table 1) *(b)* derived from fusion of Ang$^+$ P388D1 with Ang$^-$ unactivated Balb/c mouse peritoneal macrophages. Note the almost complete suppression of angiogenesis in *b* when compared to *a*.

gens which stimulate proliferation of most cells of mesodermal and neuroectodermal origin [57]. Both the acidic (aFGF) and basic (bFGF) forms of FGF were originally identified in brain tissue. Subsequent studies have revealed bFGF to be present in virtually all cells examined. FGF also appears to be synthesized by endothelial cells and may be deposited in the subendothelial extracellular matrix [58–60]. In addition to being a potent mitogen for fibroblasts, the FGFs have been shown to stimulate endothelial cell proliferation, chemotaxis and production of plasminogen activator and collagenases. Not unexpectedly, FGFs are angiogenic in vivo [61]. Studies by Baird et al. [62] have shown that macrophages also contain an FGF-like molecule(s). These workers isolated an immunoreactive FGF from thioglycolate-induced peritoneal exudate macrophages which induced proliferation of bovine aortic endothelial cells in a dose-dependent manner. These workers suggested that FGF may represent one type of MDGF. The relevance of this finding to the mechanism of macrophage-induced angiogenesis is, however, unclear. In almost all reported studies of macrophage-induced angiogenesis, angiogenic activity is readily detectable in the culture media of macrophages. In view of the fact that the FGFs lack a leader sequence necessary for cell secretion, it is not surprising that we and others have been unable to detect FGFs in conditioned media of macrophages [S.J. Leibovich et al., unpubl. observations]. It appears that under conditions in which macrophage-induced angiogenesis is normally expressed, FGF may not contribute significantly.

While the majority of macrophage angiogenic factors reported to date have been found to be mitogenic for endothelial cells, Banda et al. [63] showed that wound fluids rich in macrophages as well as conditioned media from macrophage cultures contain an angiogenic factor which is chemotactic but not mitogenic for endothelial cells. The extent to which this nonmitogenic angiogenic factor contributes to macrophage-induced angiogenic response is not yet clear. Chemotaxis of endothelial cells clearly plays an important role in angiogenic responses. Sholley et al. [64] demonstrate that significant capillary sprout formation could occur in the absence of endothelial cell proliferation; although sustained neovascularization requires both endothelial cell proliferation and migration. It is quite feasible that such a nonmitogenic angiogenic factor in concert with endothelial mitogens could contribute to angiogenic responses.

The most recent addition to the list of macrophage-derived angiogenic factors is tumor necrosis factor alpha (TNFα). TNFα is a major macrophage cytokine that was first identified in the serum of bacillus Calmette-Guérin-primed, endotoxin-treated animals, and caused hemorrhagic necrosis of tumors [65]. It is considered the principal macrophage mediator of tumor cell cytotoxicity and cytostasis in culture. TNFα is identical to cachectin which has been shown to be the major cause of wasting in chronic infectious and neoplastic diseases and is structurally and biologically related to the lymphocyte product lymphotoxin (TNFβ) [66, 67]. TNFα is a pleotrophic mediator with a broad spectrum of biological activities where it exerts both inhibitory and stimulatory effects on a variety of target cells. TNFα has several effects on human endothelial cells including induction of granulocyte/macrophage-colony stimulating factor (GM-CSF), procoagulant activity, expression of class I major histocompatibility antigens, IL-1 production, and increased adherence of blood cells [65]. It has also been reported to be cytostatic and cytotoxic for endothelial cells in vitro; it inhibits angiogenesis in vitro and induces angiogenesis in vivo [48–70]. Fràter-Schröder et al. [68] demonstrated that recombinant human TNFα reversibly inhibited growth of bovine aortic and capillary endothelial cells in culture in a dose-dependent fashion. They also showed that TNFα antagonized the growth-promoting efforts of aFGF and bFGF in a noncompetitive manner. In vivo TNFα was angiogenic in rabbit corneas and appeared to augment FGF-mediated angiogenesis. They reported that angiogenic responses induced in rabbit corneas by high concentration (>5 μg) of TNFα were associated with edema and inflammation. They concluded that TNFα induced angiogenesis via an inflammatory mechanism secondary to its cytotoxic effect. The studies of Sato et al. [69, 70] also suggest that TNFα is inhibitory for the growth of endothelial cells. These workers reported that human recombinant TNFα was both cytostatic and cytotoxic for capillary endothelials and inhibited FGF-induced capillary endothelial cell reorganization and tube formation in culture. Recent studies by Leibovich et al. [71] suggest that TNFα can induce neovascularization but in the absence of inflammation. More importantly, our studies showed that the angiogenic activity present in the conditioned media of activated macrophages is immunologically related to TNFα. We demonstrated that TNFα at concentrations ranging from 3 to 50 ng was angiogenic in rat corneas, and on the chick chorioallantoic membrane, was chemotactic for bovine adrenal gland capillary endothelial cells, and induced capillary endothelium to form tube-like structures in vitro. To determine if the angiogenic activity present in macrophage culture fluids was related to TNFα, a polyclonal antibody to TNFα was used. This antibody was found to completely neutralize the angiogenic activity present in the culture media of thioglyocolate elicited mouse peritoneal macrophages. This antibody was also found to neutralize endothelial cell chemotaxis and angiogenesis in vitro. These results indicated that TNFα is a potent mediator of angiogenesis and can do so independently of inflammation at concentrations as low as 3.5

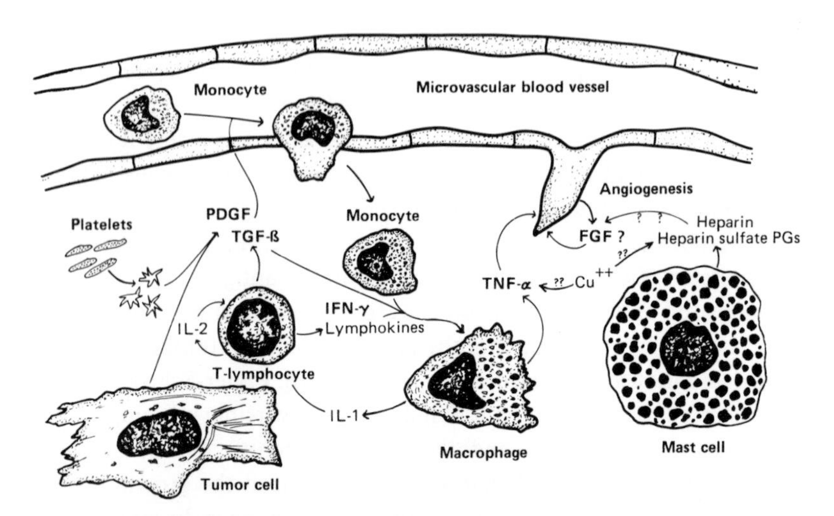

Fig. 4. Diagram [courtesy of S.J. Leibovich] depicting the possible influence of various cytokines on macrophage-induced angiogenic responses.

ng (0.2 pmol). Furthermore, it appears that a significant portion of the angiogenic activity present in macrophage-conditioned media is TNFα or an immunologically related molecule.

Conclusion

Our current interpretation of the mechanism of macrophage-induced angiogenesis as it might occur at sites of injury or during tumor development is outlined in figure 4. With the emigration of monocytes from the blood vascular system into the extravascular compartment, via the action of chemoattractants such as PDGF and TGFβ, macrophages come under the influence of potent differentiation and activation signals (i.e. TGFβ and IFN-γ). Macrophages are now capable of producing a diverse array of secretory products including some, such as TNFα, which can participate in the induction of angiogenesis. In conjunction with a number of accessory cells and mediators such as the endothelial mitogen FGF, heparin, proteolytic enzymes, chemoattractants, certain low molecular weight substances such as copper, and other mediators not shown here, the process of new blood vessel growth is initiated. These events are likely to be further amplified by cytokines secreted by accessory cells which promote the continued recruitment and activation of macrophages. Under normal conditions of wound repair, certain mediators such as TGFβ, low oxygen tension and high lactate concentration would have a positive effect during the early stages of angiogenesis. During the resolution phase the inhibitory effects of TGFβ on angiogenesis might come into play and in concert with elevated levels of oxygen, reduced lactate concentration and other inhibitors of angiogenesis to limit further neovascularization. An analogous situation might exist during tumor development. However, unlike normal wound repair, the feedback mechanisms which regulate blood vessel growth would be overwhelmed by the continuous production of tumor angiogenesis factors, and other mediators which promote the continued recruitment, differentiation and activation of macrophages.

The scenario I have outlined is by no means complete. Many of the mediators and mechanisms we and others have proposed as having a positive effect in the process of macrophage-induced angiogenesis have also been shown to have multiple and often paradoxical (inhibitory) effects. The nature of the mechanisms which orchestrate this complex and still ill-defined process will likely be the subject of intensive investigation for years to come. Since the first reports appeared implicating macrophages in the process of angiogenesis, this field has rapidly progressed to the point where it now has important implications in normal physiologic and pathologic processes. With the availability of cDNA clones and specific antibodies to many angiogenic factors, this process can now be addressed in molecular terms and eventually lead to a clarification of its mechanism of action in vivo.

Acknowledgments

I wish to thank Dr. S.J. Leibovich for reviewing the manuscript and Nancy Starks and Kathleen Stenson for their excellent stenographic services.

References

1 Folkman J, Cotran R: Relation of vascular proliferation to tumor growth. Int Rev Exp Pathol 1978;16:208–248.
2 Auerbach R: Angiogenesis-inducing factors: in Pick (ed): Lymphokines. New York, Academic Press, 1981, vol. 4, pp 69–88.
3 Folkman J: Tumor angiogenesis. Adv Cancer Res 1985;43:175–203.

4 Folkman J, Klagsbrun M: Angiogenic factors. Science 1987;235:442–447.
5 Sidkey YA, Auerbach R: Lymphocyte-induced angiogenesis (LIA): a quantitative and sensitive assay of the graft vs. host reaction. J Exp Med 1975;141:1084–1110.
6 Fromer CH, Klintworth GH: An evaluation of the role of leukocytes in the pathogenesis of experimentally induced corneal vascularization. III. Studies related to the vasoproliferative capability of polymorphonuclear leukocytes and lymphocytes. Am J Pathol 1976;82:157–168.
7 Polverini PJ, Cotran RS, Gimbrone MA Jr., et al: Activated macrophages induce vascular proliferation. Nature 1977;269:804–806.
8 Nathan C: Secretory products of macrophages. J Clin Invest 1987;79:319–326.
9 Gell PGH: Cytologic events in hypersensitivity reactions; in Lawrence (ed): Cellular and Humoral Aspects of the Hypersensitivity Studies. New York, Harper & Row, 1959, p 43.
10 Dvorak AM, Mihus MC, Dvorak HF: Morphology of delayed-type hypersensitivity reactions in man. II. Ultrastructural alterations affecting the microvasculature and the tissue mast cell. Lab Invest 1976;34:179–191.
11 Graham RC, Shannon S: Peroxidase arthritis. II. Lymphoid cell-endothelial interactions during developing immunologic inflammatory responses. Am J Pathol 1972;69:7.
12 Anderson ND, Anderson AO, Wyllie RG: Microvascular changes in lymph nodes draining skin allographs. Am J Pathol 1975;81:131–153.
13 Calderone J, Williams RT, Unanue ER: An inhibitor of cell proliferation released from cultures of macrophages. Proc Natl Acad Sci USA 1974;71:4273–4277.
14 Calderone J, Unanue ER: Two biological activities regulating cell proliferation found in cultures of peritoneal exudate cells. Nature 1975;253:359–362.
15 Calderone J, Kiely JM, Lefka J, et al: The modulation of lymphocyte functions by molecules secreted by macrophages. I. Description and partial biochemical analysis. J Exp Med 1976;142:151–160.
16 Unanue ER, Kiely JM, Calderone J: The modulation of lymphocyte functions by molecules secreted by macrophages. II. Conditions leading to increased secretion. J Exp Med 1976;144:155–150.
17 Leibovich SJ, Ross R: The role of macrophages in wound repair. A study with hydrocortisone and antimacrophage serum. Am J Pathol 1975;78:71–100.
18 Leibovich SJ, Ross R: A macrophage-dependent factor that stimulates the proliferation of fibroblasts in vitro. Am J Pathol 1976;84:501–513.
19 Polverini PJ, Cotran RA, Sholley MM: Endothelial proliferation in the delayed hypersensitivity reaction: An autoradiographic study. J Immunol 1977;118:529–532.
20 Gillespie GY, Estes JE, Pledger WJ: Macrophage-derived growth factors for mesenchymal cells; in Pick (ed): Lymphokines. New York, Academic Press, 1986, vol 11, pp 213–242.
21 Clark RA, Stone RD, Leung DK, et al: Role of macrophages in wound healing. Surg Forum 1976;27:16–18.
22 Thakral KK, Goodson WH, Hunt TK: Stimulation of wound blood vessel growth by wound macrophages. J Surg Res 1979;26:430–436.
23 Hunt TK, Knighton DR, Thakral KK, et al: Studies on inflammation and wound healing: angiogenesis and collagen synthesis stimulated in vivo by resident and activated wound macrophages. Surgery 1984;96:48–54.

24 Moore JW III, Sholley MM: Comparison of the neovascular effects of stimulated macrophages and neutrophils in autologous rabbit corneas. Am J Pathol 1985;120: 87–98.
25 Evans R: Effect of X-irradiation on host cell infiltration and growth of murine fibrosarcoma. Br J Cancer 1977;35:557–566.
26 Evans R: The effect of azathioprine on host-cell infiltration and growth of a murine fibrosarcoma. Int J Cancer 1977;20:120–128.
27 Mostafa LK, Jones DB, Wright DH: Mechanism of the induction of angiogenesis by human neoplastic lymphoid tissue: studies on the chorioallantoic (CAM) of the chick embryo. J Pathol 1980;132:91–205.
28 Mostafa LK, Jones DB, Wright DH: Mechanism of the induction of angiogenesis by human neoplastic lymphoid tissue: studies employing bovine aortic endothelial cells in vitro. J Pathol 1980; 132:207–216.
29 Stenzinger W, Brüggen J, Macher E, et al: Tumor angiogenic activity (TAA) production in vitro and growth in the nude mouse by human malignant melanoma. Eur J Cancer Clin Oncol 1982;19:649–656.
30 Polverini PJ, Leibovich SJ: Induction of neovascularization in vivo and endothelial proliferation in vitro by tumor associated macrophages. Lab Invest 1984;51: 635–642.
31 Polverini PJ, Leibovich SJ: Effect of macrophage depletion on growth and neovascularization of hamster buccal pouch carcinomas. J Oral Pathol 1987;16:436–441.
32 Brown RA, Weiss JB, Tomlinson IW, et al: Angiogenesis factor from synovial fluid resembling that from tumors. Lancet 1980;i:682–685.
33 Brown RA, Tomlinson JW, Hill CR, et al: Relationship of angiogenesis factor in synovial fluid to various joint diseases. Ann Rheum Dis 1983;42:301–307.
34 Semble EL, Turner RA, McCrickard EL: Rheumatoid arthritis and osteoarthritis synovial fluid effects on primary human endothelial cell cultures. J Rheumatol 1985;12:237–241.
35 Harris ED Jr: Recent insights into the pathogenesis of the proliferative lesions in rheumatoid arthritis. Arthritis Rheum 1976;19:68–72.
36 Krane SM, Golding SR, Dayer JM: Interactions among lymphocytes, monocytes and other synovial cells in the rheumatoid synovium. Lymphokines 1982;7:75–136.
37 Koch AE, Polverini PJG, Leibovich SJ: Stimulation of neovascularization by human rheumatoid synovial tissue macrophages. Arthritis Rheum 1986;29:471–479.
38 Adams DG, Hamilton TA: The cell biology of macrophage activation. Ann Rev Immunol 1984;3:283–310.
39 Koch AE, Polverini PJ, Leibovich SJ: Induction of neovascularization by activated human monocytes. J Leukocyte Biol 1986;39:233–238.
40 Polverini PJ, Leibovich SJ: Induction of neovascularization and nonlymphoid mesenchymal cell proliferation by macrophage cell lines. J Leukocyte Biol 1985; 37:279–288.
41 Knighton DR, Hunt TK, Scheuenstuhl H, et al: Oxygen tension regulates the expression of angiogenesis factor by macrophages. Science 1983;221:1283–1285.
42 Jensen JA, Hunt TK, Scheuenstuhl H, et al: Effect of lactate, pyruvate, and pH on secretion of angiogenesis and mitogenesis factors by macrophages. Lab Invest 1986;54:574–578.
43 Roberts AB, Sporn MB: Transforming growth factors. Cancer Surveys 1985;4:683–705.

44 Roberts AB, Anzano MA, Wakefield LM, et al: Type β transforming growth factor: a bifunctional regulator of cellular growth. Proc Natl Acad Sci USA 1985;82: 119–123.
45 Sporn MB, Roberts AB, Wakefield LW, et al: Transforming growth factor-β: biology function and chemical structure. Science 1986;233:532–534.
46 Sporn MB, Roberts AB, Shull JH, et al: Polypeptide transforming growth factors isolated from bovine sources and used for wound healing in vivo. Science 1983;219:1329–1331.
47 Roberts AB, Sporn MB, Assoian RK, et al: Transforming growth factor type β: rapid induction of fibrosis and angiogenesis in vivo and stimulation of collagen formation in vitro. Proc Natl Acad Sci USA 1986;83:417–471.
48 Mustoe TA, Pierre GF, Thomason A, et al: Accelerated healing of incisional wounds in rats induced by transforming growth factor-β. Science 1987;237:333–336.
49 Wahl SM, Hunt DA, Wakefield LM, et al: Transforming growth factor type β induces monocyte chemotaxis and growth factor production. Proc Natl Acad Sci USA 1987;84:5788–5792.
50 Assoian RK, Sporn MB: Type-β transforming growth factor in human platelets: release during platelet degranulation action on vascular smooth muscle. J Cell Biol 1986;1023:1217–1223.
51 DeLarco JE,, Todaro GJ: Growth factors from murine sarcoma virus-transformed cells. Proc Natl Acad Sci USA 1978;75:4001–4005.
52 Roberts AB, Frolik CA, Azano MA, et al: Transforming growth factors from neoplastic and nonneoplastic tissues. Fed Proc 1983;42:2621–2626.
53 Fràter-Schröder M, Müller F, Birchmeier W, et al: Transforming growth factor-beta inhibits endothelial cell proliferation. Biochem Biophys Res Commun 1986; 137:295–302.
54 Heimark RL, Twardzik DR, Schwartz SM: Inhibition of endothelial regeneration by type-beta transforming growth factor from platelets. Science 1986;233:1078–1080.
55 Bouck N, Stoler A, Polverini PJ: Coordinate control of anchorage independence, actin cytoskeleton, and angiogenesis by human chromosome 1 in hamster-human hybrids. Cancer Res 1986;46:5101–5105.
56 Shimakado K, Raines EW, Madtes DK, et al: A significant part of macrophage-derived growth factor consists of at least two forms of PDGF. Cell 1985;43:277–286.
57 Thomas KA: Fibroblast growth factors. FASEB J 1987;1:434–440.
58 Schweigerer L, Neufeld G, Friedman J, et al: Capillary endothelial cells express basic fibroblast growth factor, a mitogen that promotes their own growth. Nature 1987; 325:257–259.
59 Voldavsky I, Folkman J, Sullivan R, et al: Endothelial cell-derived basic fibroblast growth factor: synthesis and deposition into subendothelial extracellular matrix. Proc Natl Acad Sci USA 1987;84:2292–2296.
60 Baird A, Ling N: Fibroblast growth factors are present in the extracellular matrix produced by endothelial cells in vitro: implications for a role of heparinase-like enzymes in the neovascular response. Biochem Biophys Res Commun 1987; 142:428–435.
61 Thomas KA, Gimenez-Gallego G: Fibroblast growth factors; broad spectrum mitogens with potent angiogenic activity. Trends Biochem Sci 1986;11:1–4.

62 Baird A, Marmède P, Böhlen P: Immunoreactive fibroblast growth factor in cells of peritoneal exudate suggests its identity with macrophage-derived growth factor. Biochem Biophys Res Commun 1985;126:358–364.
63 Banda MJ, Knighton DR, Hunt TK, et al: Isolation of a nonmitogenic angiogenesis factor from wound fluid. Proc Natl Acad Sci USA 1982;79:7773–7777.
64 Sholley MM, Ferguson GP, Sibel H, et al: Mechanisms of neovascularization. Vascular sprouting can occur without proliferation of endothelial cells. Lab Invest 1984;51:624–634.
65 Le J, Vilćek J: Biology of disease. Tumor necrosis factor and interleukin-1: cytokines with multiple overlapping biological activities. Lab Invest 1987;56:234–248.
66 Beutler B, Greenwald D, Hulmes JD, et al: Identity of tumor necrosis factor and the macrophage-secreted factor cachectin. Nature 1985;316:552–554.
67 Pennica D, Nedwin GE, Hayflick JS, et al: Human tumor necrosis factor: precursor structure, expression and homology to lymphotoxin. Nature 1984;312:724–729.
68 Fràter-Schröder M, Risau W, Hillman R, et al: Tumor necrosis factor type α, a potent inhibitor of endothelial cell growth in vitro, is angiogenic in vivo. Proc Natl Acad Sci USA 1987;84:5277–5281.
69 Sato N, Sawasaki S, Haranaka K: Cytotoxic effects of rabbit tumor necrosis factor against capillary endothelial cells. Proc Jap Soc Immunol 1985;15:178.
70 Sato N, Fukuda K, Nariuchi H, et al: Tumor necrosis factor inhibiting angiogenesis in vitro. J Natl Cancer Inst 1987;79:1383–1391.
71 Leibovich SJ, Polverini PJ, Shepard HM, et al: Macrophage-induced angiogenesis is mediated by tumour necrosis factor α. Nature 1987;329:630–632.

Dr. Peter J. Polverini, Department of Pathology, Northwestern University
Medical and Dental Schools, 303 East Chicago Avenue, Chicago, IL 60611 (USA)

Sorg C (ed): Macrophage-Derived Cell Regulatory Factors.
Cytokines. Basel, Karger, 1989, vol 1, pp 74–88

Biology of Cachectin[1]

Yuman Fong[a], *Kevin J. Tracey*[a,b], *Stephen F. Lowry*[a], *Anthony Cerami*[b]

[a]Laboratory of Surgical Metabolism, New York Hospital-Cornell Medical Center, and [b]Laboratory of Medical Biochemistry, Rockefeller University, New York, N.Y., USA

Introduction

The deleterious changes associated with infection and injury are triggered by a cascade of endogenous hormones and cytokine mediators. The cytokine 'cachectin' appears to play a pivotal role in this host response to injury. Isolated for its effects on lipoprotein metabolism, and independently for its ability to produce necrosis of certain tumor lines, this macrophage-secreted polypeptide is a pluripotent mediator capable of directly eliciting many of the tissue responses to injury and inflammation. It occupies a central role in propagating and amplifying the cascade of endogenous mediators seen in injury or infection. Moreover, cachectin appears to be capable of triggering the diverse clinical syndromes of cachexia and shock.

History of Cachectin

Cachectin was first identified during the study of weight loss and tissue wasting accompanying chronic disease. Rabbits infected with the parasitic disease *Trypanosoma brucei* develop severe cachexia, losing up to 50% of lean body mass within weeks [1]. A progressive hypertriglyceridemia also occurs, which was accounted for by a defect in systemic lipoprotein lipase (LPL) activity [1]. Since these metabolic changes occurred even in the

[1] Supported in part by grants numbers KO4GM-00505 and AI21359 from the National Institute of Health, Bethesda, Md. Y.F. is supported by a Clinical Fellowship from the American Cancer Society.

presence of a relatively low parasite load, Kawakami and Cerami [2] postulated that a circulating factor produced by the host was responsible.

Hypertriglyceridemia and suppression of LPL activity was also noted after lipopolysaccharide (LPS) injection in endotoxin-sensitive (C3H/HeN) mice [2]. The presence of a transferable serum factor in these endotoxin-challenged C3H/HeN mice was noted that could elicit similar alterations in lipid metabolism in (C3H/HeJ) mice which are normally resistant to the hyperlipidemic effects of endotoxin. This factor was named 'cachectin' for its presumed role as a mediator of altered cellular energy metabolism and cachexia [3].

Cachectin was subsequently isolated and purified as a secretory product of activated murine macrophages (RAW 264.7). Since this factor is a potent suppressor of LPL in the adipocyte cell line 3T3-L1, Beutler et al. [3] used this biological assay to purify cachectin to homogeneity. The amino terminal sequence of cachectin was noted to have great homology to the monokine tumor necrosis factor alpha, previously identified for its ability to produce necrosis of certain tumor lines [4]. Subsequent immunological and genetic sequence analysis have demonstrated that cachectin and tumor necrosis factor are identical [5–7].

Physical Properties of Cachectin

Cachectin is a polypeptide hormone with a molecular weight of 17,000. The structure of this protein from rabbit, mouse, and human sources has been determined [6–8]. A remarkable interspecies conservation exists; 79% of the amino acid residues are conserved between the mouse and human cachectin [7, 8].

Human cachectin is translated as a 233 amino acid prohormone which is then cleaved during biological processing to form the 157 amino acid active cachectin protein [9]. The 76 additional amino acids in the prohormone sequence are attached to the N-terminus of the mature protein and biologic activity of this sequence has yet to be determined. The mature protein exists as a dimer, trimer, or pentamer in solution [10]. The human protein has a pI of approximately 5.3, and is a relatively hydrophobic protein containing one intrachain disulfide bond [11]. It is irreversibly denatured by boiling and also loses some biological activity upon freezing and thawing.

Specific receptors for cachectin have been found in a wide variety of cell types [12–15]. Cultured adipocytes 3T3-L1 and C2 myotubules have been shown to have approximately 10,000 receptors per cell, with a dissociation constant (kd) of 3×10^{-9} *M* [3]. Murine fibrosarcoma L929 cells, which are sensitive to the cytotoxic effects of cachectin, have 2,200 receptors per cell with a kd of 6.1×10^{-10} *M* [15], while cachectin-resistant fibroblasts FS-4 are known to have 7,500 receptors per cell with a kd of 3.2×10^{-10} *M*. Therefore, the existence or number of receptors on the cell do not correlate to the biological responsiveness of a cell type to cachectin. As is true for many circulating hormones, a maximal biological response can be elicited by occupancy of as few as 5% of the receptors by cachectin [16]. The postreceptor mechanisms triggering the biological responses are incompletely understood, but in many cells biological responses are observed after cellular uptake of the molecule [13].

Production/Excretion

The gene for human cachectin is located on the short arm of chromosome 6, in close proximity to lymphotoxin, another immunomodulator with which cachectin shares significant amino acid sequence homology [17]. In the mouse, the cachectin gene is located on chromosome 17, in close proximity to the D locus of the major histocompatibility complex [10]. Expression of cachectin is tightly controlled, both at the transcriptional and translational level [18]. The human cachectin gene contains three introns. In the 3′-untranslated region of the cachectin mRNA is a repeating octameric unit composed solely of adenosine and uridine residues [6]. Many other inflammatory mediators, such as interleukin-1 (IL-1), lymphotoxin, granulocyte macrophage colony-stimulating factor (GM-CSF), and interferons, also contain repeating sequences of this octameric unit (UUAUUUAU)n [6]. Insertion of this sequence confers destabilization to the messenger RNA for the globin protein, reducing the half-life of this normally stable message [19]. Thus, this sequence may be involved in the posttranscriptional regulation of cachectin production.

A variety of cells derived from the myeloid line, such as blood monocytes, pulmonary macrophages, liver Kupffer cells [20], or peritoneal macrophages [21], are capable of producing cachectin. Mast cells and natural killer cells have also been shown to synthesize this protein [21]. Unstimulated monocytes express low levels of cachectin mRNA. Stimulation of the monocytes then induces both increased transcription and increased translation for the protein [18], leading to the appearance of mature protein within minutes.

Secretion of cachectin is elicited by a large assortment of infectious or inflammatory stimuli including bacteria or the bacterial cell wall-derived LPS [18]. Membranes of the parasites *T. brucei* and *Plasmodium* will also elicit release of cachectin/TNF by peritoneal exudate cells [22]. Additionally, viral particles increase transcription and translation of cachectin by blood monocytes [23]. *Mycobacterium tuberculosis* can also trigger release of cachectin from human monocytes [24]. Secretion of cachectin is inhibited at the transcriptional and translational levels by prophylactic administrations of glucocorticoids [25].

The half-life of circulating cachectin is brief. Radioiodinated cachectin injected into mice is cleared with a half-life of approximately 6 min and this agrees closely with data obtained in rabbits [26]. In humans, the circulating cachectin levels after intravenous infusions can be described with a monoexponential equation with a half-life of 14–18 min [27]. After the injection of radioiodinated cachectin into animals, label was detected in many organ systems, including 21% in the liver, 30% in the skin, 9% in the gastrointestinal tract, and 8% in the kidney [3]. Significant radioactivity without biological activity is also found in the urine, suggesting that the molecule is quickly degraded. While the precise tissue distribution is incompletely characterized, of note is that little radioactivity is found in the central nervous system, suggesting that circulating cachectin may not effectively cross the blood-brain barrier.

Cachectin is detected early after bacterial or LPS infusions in animals. After such infusions, a monophasic cachectin appearance in the circulation is noted. In rats, rabbits [26], and baboons [28], circulating cachectin levels rise within 30 min, peak 90–120 min after the infusion, and return to baseline levels after approximately 4 h. Bolus LPS infusions in man also demonstrate a similar monophasic pattern, with circulating levels peaking approximately 1.5 h after LPS injection [28].

Tissue Effects of Cachectin

The tissue effects of cachectin have been extensively reviewed [9, 10, 21, 29]. Many tissues, including adipocytes, myocytes, macrophages, and osteocytes, are known to possess receptors for cachectin. In vitro exposure of these cell and tissue types to cachectin results in many cellular changes commonly seen in injury and catabolic illness. Thus, cachectin has been implicated as a cause of the pathophysiologic changes seen in injury and illness.

Skeletal Muscle

Incubation of skeletal muscle with cachectin produces a reduction of the skeletal muscle resting transmembrane potential [30]. A similar reduction of skeletal muscle resting membrane potential is noted in vivo in canines following cachectin administration [31]. This change in membrane function resembles that seen in septic or thermally injured patients and may represent a primary pathophysiologic mechanism underlying the sequestration of extravascular water and sodium seen in sepsis and shock [32].

Cachectin also appears to directly influence skeletal muscle carbohydrate metabolism. In L-6 myotubules, this protein increases transcription of the hexose transporter, increases transport of glucose, depletes cell glycogen, and increases cellular efflux of lactate [33]. Thus, cachectin may represent a signal for the early induction of anaerobic glycolysis in somatic tissues following injury.

Adipocytes

Cachectin produces depletion of lipid stores in adipocytes. Incubation of adipocytes with cachectin reduces the activity of LPL, the membrane-bound protein primarily responsible for the clearance of extracellular lipids by adipocytes [2]. This LPL suppression appears to be regulated at the transcriptional level [34]. Cachectin also suppresses fatty acid synthesis: incubation of 3T3-L1 adipocytes with cachectin produces a decreased uptake of acetate [35], and a decreased incorporation of labeled glucose [29] into fatty acids. Cachectin also stimulates lipolysis directly by stimulation of the hormone-sensitive lipases. The decreased clearance of extracellular lipids, inhibition of fatty acid synthesis, and increased cellular lipolysis result in net loss of triglycerides from adipocytes. This loss in cellular lipid content is accompanied by a morphologic reversion of adipocytes to fibroblasts [36].

Injection of cachectin increases hepatic production of triglycerides [37] in rodents. Two hours after cachectin infusion, a rise in serum triglycerides is accompanied by increased hepatic lipogenesis as measured by incorporation of tritiated water into fatty acids [37]. Additionally, the rise in serum triglyceride levels observed in primates infused with gram-negative bacteria is blocked by prophylactic treatment with monoclonal antibody specific for cachectin [Fong et al., in preparation]. In concert, the alterations in lipid metabolism triggered by cachectin may provide a mechanism for the hypertriglyceridemia, and the exaggerated loss of body fat seen in infection and other catabolic states.

Liver

Cachectin administration produces accumulation of liver mass and alterations in hepatic protein metabolism. Hepatomegaly can be noted as early as 17 h after an infusion of cachectin in rats [37]. The change in hepatic mass is at least partly related to a synergistic influence upon glucagon-mediated increases in hepatic uptake of amino acids [38], and occurs despite the total body weight loss that is seen with chronic cachectin administration. Hepatic morphologic changes occur with chronic cachectin administration including bile duct proliferation, infiltration by monocytes, and periportal inflammatory changes [39].

Cachectin also produces functional changes in liver protein synthesis. In hepatocyte cell lines, cachectin increases the transcription and appearance of certain acute phase proteins while decreasing the production of albumin [40]. In mice, administration of cachectin is associated with an increase in the circulating levels of acute phase proteins [41] and an accumulation of mRNA for certain acute phase proteins in the liver. These hepatic cellular alterations are reminiscent of the hepatomegaly, acute phase protein response, and increase in hepatic protein synthesis seen after injury.

Endothelium

Cachectin produces many morphologic and functional changes in vascular endothelial cells. This polypeptide produces morphologic reorganization of human vascular endothelial cell monolayers, causing cell elongation, overlapping, and alterations in the cytoskeleton [42]. Incubation of human umbilical vein endothelial cells with cachectin produces a dose- and time-dependent increase in neutrophil adherence through the expression of cell surface antigens [43]. Certain endothelial cell surface antigens that are observed in vivo in delayed-type sensitivity reactions, can be elicited by cachectin in vitro [44]. Incubation with cachectin also increases procoagulant activity on the endothelial surface [45], and vascular permeability [46]. Additionally, recent evidence suggests cachectin to be a growth factor, possessing not only angiogenic potential [47], but also potential to stimulate fibroblast proliferation [48]. With such capacities, cachectin may facilitate localization of injury and promote an inflammatory response.

Leukocytes

Cachectin stimulates neutrophil release from the bone marrow, as well as enhancing neutrophil margination and activation [49]. Cachectin induces neutrophil degranulation, superoxide production, and lysozyme release [50].

This activation of neutrophils enhances antibody-dependent cellular cytotoxicity and neutrophil-mediated inhibition of fungal growth [51].

Cachectin also promotes differentiation of myelogenous cells to monocyte/macrophages [52], and activates macrophages [53]. IL-1 release from macrophages and endothelial cells is also enhanced. Activation of macrophages participates in the inhibition of intracellular replication of viral and parasitic organisms [54], and increases the cytotoxicity against virally infected cells [55].

Interaction with Other Mediators

The biological importance of cachectin results not only from the direct effect of this protein on target organs, but also from the interactions of this protein with other locally produced and circulating mediators. Cachectin appears to hold a very proximal position in the cascade of endogenous humoral responses to injury and infection, and therefore is important in triggering the release of counterregulatory hormones and other immunomodulators at the time of injury. Once released, the actions of cachectin also appear to be synergistic with those of many other mediators.

Cachectin clearly stimulates secretion of the classical counterregulatory hormones. Infusion of cachectin in canines is known to elicit an increase in circulating levels of glucagon, cortisol, and the catecholamines [31]. Prophylactic administration of monoclonal antibodies to cachectin to primates receiving live *Escherichia coli* attenuates the normal elevation of such hormones [56].

In murine resident peritoneal macrophages, cachectin stimulates release of IL-1 and prostaglandin E_2 (PGE_2) [57]. Cachectin enhances transcription and translation of interferon-β_2 in human fibroblasts [58]. Furthermore, cachectin not only appears to induce production of other mediators, such as GM-CSF or platelet-activating factor (PAF), but also stimulates production of more cachectin [59]. IL-1 also appears capable of inducing cachectin release [60]. Thus, cachectin holds a pivotal role in the monokine/counterregulatory hormone cascade that once triggered, feeds back positively upon itself, and produces many of the clinical findings we associate with critical illness.

Not only does cachectin trigger the release of many other mediators, this protein also acts synergistically with many other humoral mediators including glucagon, IL-1, and interferon-γ. Recombinant human cachectin enhances the glucagon-mediated transport of amino acids into the liver [61]. The synergistic action of cachectin with IL-1 in many tissue types including pituitary cells, bone, vascular endothelial cells, skin, fibroblasts, and the islets of Langerhans is well described [62]. Cachectin is synergistic with interferon-γ in their antiproliferative effects on certain human and murine cell lines [48] and in their enhancement of fungal killing by polymorphonuclear cells [51]. The mechanism of this synergism may involve an increase in receptor binding. The binding of cachectin to adipocytes and fibroblasts is known to be enhanced by preexposure of the cells to interferon-γ [33].

LPS itself may potentiate the effects of cachectin, either directly or by inducing release of other mediators. In endotoxin-sensitive mice, endotoxin greatly enhanced the ability of cachectin to produce death or hemorrhagic necrosis [63]. In addition, lysates of *Mycoplasma*-infected cells are also synergistic with cachectin in producing hemorrhagic necrosis [63].

Systemic Effects of Cachectin

Cachectin infusions produce many of the systemic findings associated with disease and injury. In man, for example, infusion of cachectin produces a rise in the counterregulatory hormones, an increase in C-reactive protein, a decrease in zinc levels, fever, and an efflux of amino acids from the extremity [64]. Cachectin infusions into animals produce two particular syndromes, cachexia and shock, that bear such close resemblance to the chronic cachexia of disease and to septic shock that cachectin has been implicated as a major causative factor in these clinical syndromes.

Cachexia

Chronic diseases, such as malignancies, parasitic diseases, and tuberculosis, frequently lead to a syndrome of cachexia characterized by anorexia, tissue wasting, weight loss and anemia [65]. Cachexia-associated wasting may be a primary determinant of poor clinical outcome and death. Although widely studied and very common, the mechanisms and mediators responsible for this cachexia of chronic disease are incompletely understood.

Cachectin was first implicated as a mediator of cachexia in studies employing the injection of macrophage supernatants in mice [66]. With the availability of recombinant cachectin, chronic administration of sublethal doses of this protein in experimental models reproduced many of the physiologic findings observed in the chronic wasting of clinical disease. In rodents, such a regimen produces a decrease in food intake [39, 41], decreased nitrogen balance [67], lipid depletion, and weight loss. These metabolic changes are alleviated by simultaneous administration of antibody to cachectin [39]. Tumors secreting recombinant human cachectin implanted into nude mice also produce a similar syndrome of severe weight loss and cachexia [68], which adversely influences survival. Of note is that animals appear to become tolerant to repeated intraperitoneal administration of human cachectin, requiring increasing doses to elicit persistent anorexia [39]; this tolerance was not reported in either intravenous administration [67] or in a study utilizing cachectin-secreting tumors [68]. The mechanisms underlying this phenomenon are currently under investigation.

The exaggerated loss of skeletal muscle protein during in vivo chronic administration of cachectin [39] may be triggered by secondary hormones or cytokines, since no direct effect of cachectin on skeletal muscle protein metabolism has been demonstrated. Indeed, incubation of isolated murine extensor digitorum longus with cachectin does not appear to alter muscle protein balance or degradation [69]. Cortisol is a possible candidate as a secondary mediator of this wasting effect, since cachectin produces a decrease in skeletal muscle uptake of the alanine analog aminoisobutyrate in normal rats but not in adrenalectomized rats [70].

Chronic cachectin administration also produces a reduction in body red blood cell mass reminiscent of the anemia of chronic disease. Tracey et al. [39] noted that anemia in rats treated with cachectin was associated with a decrease in red cell mass.

Shock

Severe infection or endotoxemia frequently progress to overwhelming cardiovascular collapse, shock, and death. This syndrome of shock is characterized by hypotension, reduced cardiac output, decreased tissue perfusion leading to organ failure, and lactic acidosis [71].

High dose cachectin administration in animals is capable of precipitating a syndrome similar to that seen in human septic shock. Acute infusion in rats produces hypotension, lactic acidosis, and death. There is also hemoconcentration, suggesting extravascular sequestration of fluids. Necropsy reveals adrenal necrosis, pulmonary congestion, cecal necrosis and ischemia of other regions of the bowel: findings not unlike those seen during septic shock in man [72].

Acute infusions of cachectin in canines also produce cardiovascular collapse with progressive hypotension, decreasing cardiac output, and death

[31]. There is also an associated rise in circulating counterregulatory hormones, release of lactate from peripheral tissues, and a depression of the skeletal muscle resting membrane potential. There is a sequestration of fluids extravascularly, as evidenced by increased fluid requirements following the cachectin infusion.

The most convincing data on the role of cachectin in the pathogenesis of endotoxic shock comes from studies utilizing blockade of cachectin during bacterial and LPS infusions. Prophylactic administration of a rabbit antiserum to cachectin conferred survival benefit in a dose-dependent fashion to mice subsequently treated with endotoxin [73]. Tracey et al. [56] also protected baboons against the lethal consequences, and many of the metabolic consequences, of severe bacteremia by prophylactically immunizing these primates with monoclonal antibodies to cachectin.

Circulating Levels in Patients

Not only can cachectin mimic many of the pathophysiologic findings of disease, a higher frequency of circulating cachectin can be found in patients suffering a variety of disease states. Almost 70% of patients afflicted with the parasitic diseases leishmaniasis and malaria were found to have detectable levels [74]. In patients with meningococcal infection, circulating cachectin was detected with a significantly greater frequency than in healthy controls, and levels greater than 0.1 ng/ml correlated with increased mortality [75]. Detectable levels of cachectin have been reported in as many as 50% of patients with cancer [76, 77]. Higher frequencies of detectable cachectin levels have also been described in other conditions such as thermal injury [78] and renal allograft rejection [79].

Caution must be exercised in interpreting the absolute circulating levels of this protein for a number of reasons: (1) Available assays are only sensitive to approximately 30 pg/ml, but levels of cachectin below this are biologically active. (2) The transient nature of cachectin appearance, as shown in animal models, renders the interpretation of random blood sampling during chronic disease difficult. (3) Cachectin is released by cells at a number of tissue sites, including Kupffer cells in the liver, pulmonary macrophages, peritoneal macrophages, and glial cells. Cachectin and other cytokines likely exert important paracrine effects in addition to transient endocrine effects. Therefore, tissue concentrations of this protein may be biologically more significant than that which is intermittently detected in the circulation. (4)

Cachectin is also synergistic with many other endogenous mediators in biologic activity, especially IL-1 and interferon-γ, which are also released during inflammation and disease. Thus, the net biologic effect of cachectin depends not only on the local or circulating levels of cachectin, but also on the coincident levels of these other synergistic mediators.

Conclusions

Cachectin is a monocyte product with a myriad of potent biologic actions. In small amounts this protein may confer benefit to organisms subjected to injury. The stimulation of macrophages or neutrophils by cachectin may be important in the body's immune reaction to infection and parasitic disease. The stimulation of vascular and fibroblast proliferation by cachectin may function in wound healing, or in improving abscess formation: a process vital in containing life-threatening infections. A prolonged or massive production of this protein may, however, lead to severe metabolic and hemodynamic derangements, or even shock. Monoclonal antibodies and antiserum to cachectin have been successful at blocking many of these deleterious effects and may be useful as clinical therapy.

References

1 Rouzer CA, Cerami A: Hypertriglyceridemia associated with *Trypanosoma brucei* infections in rabbits: role of effective triglyceride removal. Mol Biochem Parasitol 1980;2:31–38.
2 Kawakami M, Cerami A: Studies of endotoxin-induced decrease in lipoprotein lipase activity. J Exp Med 1981;154:631–639.
3 Beutler B, Mahoney J, Le Trang N, et al: Purification of cachectin, a lipoprotein lipase-suppressing hormone secreted by endotoxin-induced RAW 264.7 cells. J Exp Med 1985;161:984–995.
4 Carswell EA, Old LJ, Kassel RL, et al: An endotoxin-induced serum factor that causes necrosis of tumors. Proc Natl Acad Sci USA 1975;72:3666–3670.
5 Beutler B, Greenwald D, Hulmes KD, et al: Identity of tumor necrosis factor and the macrophage secreted factor cachectin. Nature 1985;316:552–554.
6 Caput D, Beutler B, Hartog K, et al: Identification of a common nucleotide sequence in the 3′-untranslated region or mRNA molecules specifying inflammatory mediators. Proc Natl Acad Sci USA 1986;83:1670–1674.
7 Fransen L, Muller R, Marmenout, et al: Molecular cloning of mouse tumor necrosis factor cDNA and its eukaryotic expression. Nucleic Acids Res 1985;13:4417–4429.
8 Pennica D, Nedium GE, Hayflick JS, et al: Human tumor necrosis factor: precursor structure, expression and homology to lymphotoxin. Nature 1984;312:724–729.

9 Beutler B, Cerami A: Cachectin: more than a tumor necrosis factor. New Engl J Med 1987;316:379–385.
10 Beutler B, Cerami A: Cachectin and tumor necrosis factor as two sides of the same biologic coin. Nature 1986;320:584–588.
11 Aggarwal BB, Kohr WJ, Hass PE, et al: Human tumor necrosis factor, production, purification, and characterization. J Biol Chem 1985;260:2345–2354.
12 Aggarwal BB, Eessalu TE, Hass PE: Characterization of receptors for human tumor necrosis factor and their regulation by gamma interferon. Nature 1985;318:665–667.
13 Baglioni C, McCandless S, Tavernier J, et al: Binding of human tumor necrosis factor to high affinity receptors on HeLa and lymphoblastoid cells sensitive to growth inhibition. J Biol Chem 1985;260:13395–13397.
14 Kull FC, Jacobs S, Cuatrecasas P: Cellular receptor for ^{125}I-labeled tumor necrosis factor: specific binding, affinity labeling, and relationship to sensitivity. Proc Natl Acad Sci USA 1985;82:5756–5760.
15 Tsujimoto M, Yip YK, Vilcek J: Tumor necrosis factor: specific binding and internalization in sensitive and resistant cells. Proc Natl Acad Sci USA 1985; 82:7626–7630.
16 Tsujimoto M, Vilcek J: Tumor necrosis factor receptors in HeLa cells and their regulation by interferon gamma. J Biol Chem 1986;261:5384–5388.
17 Nedwin GE, Naylor SL, Sakaguchi AY, et al: Human lymphotoxin and tumor necrosis factor genes structure, homology and chromosomal localization. Nucleic Acids Res 1985;13:6361–6371.
18 Beutler B, Krochin N, Milsark IW, et al: Control of cachectin (tumor necrosis factor) synthesis: mechanism of endotoxin resistance. Science 1986;232:977–980.
19 Shaw G, Kamen R: A conserved AU sequence from the 3′-untranslated region of GM-CSF mRNA mediated selective mRNA degradation. Cell 1986;46:659–667.
20 Hesse DG, Davatelis G, Felsen D, et al: Cachectin/tumor necrosis factor gene expression in Kupffer cells. J Leukocyte Biol 1987;42:422.
21 Beutler B, Cerami A: Cachectin: more than a tumor necrosis factor. New Engl J Med 1987;316:379–385.
22 Hotez PJ, LeTrang N, Fairlamb AH, et al: Lipoprotein lipase suppression in 3T3-L12 cells by a hematoprotozoan induced mediator from peritoneal exudate cells. Parasite Immunol 1984;6:203–209.
23 Wong GHW, Goeddal DV: Tumor necrosis factor alpha and beta inhibit virus replication and synergize with interferons. Nature 1986;23:819–821.
24 Rook GAW, Paverne J, Leveton C, et al: The role of gamma-interferon, vitamin D_3 metabolites, and tumor necrosis factor in the pathogenesis of tuberculosis. Immunology 1987;62:229–234.
25 Beutler B, Milsark IW, Krochin N, et al: Cachectin (tumor necrosis factor, TNF): the monokine responsible for the lethal effects of endotoxemia. Blood 1985; 66:83a.
26 Beutler BA, Milsark IW, Cerami A: Cachectin/tumor necrosis factor: production, distribution and metabolic fate in vivo. J Immunol 1985;35:3972–3977.
27 Blick M, Sherwin SA, Rosenblum M, et al: Phase I study of recombinant tumor necrosis factor in cancer patients. Cancer Res 1987;47:2986–2989.
28 Hesse DG, Tracey KJ, Fong Y, et al: Cytokine appearance in human endotoxemia and non-human primate bacteremia. Surg Gynecol Obstet 1988;166:147–153.

29 Tracey KJ, Lowry SF, Cerami A: Physiological responses to cachectin: in Tumor Necrosis Factor and Related Cytotoxins. Ciba Found Symp 131. Chichester, Wiley, 1987, pp 88–108.
30 Tracey KJ, Lowry SF, Beutler B, et al: Cachectin/tumor necrosis factor mediates changes of skeletal muscle plasma membrane potential. J Exp Med 1986;164:1368–1373.
31 Tracey KJ, Lowry SF, Fahey TJ, et al: Cachectin/tumor necrosis factor induces lethal septic shock and stress hormone responses in the dog. Surg Gynecol Obstet 1987;164:415–422.
32 Illner HP, Shires GT: Membrane defect and energy status of rabbit skeletal muscle cells in sepsis and shock. Arch Surg 1981;116:1302–1307.
33 Lee DM, Zentella A, Pekala PH, et al: Effect of endotoxin induced monokines on glucose metabolism in the muscle cell line L-6. Proc Natl Acad Sci USA 1987;84:2590–2594.
34 Zechner R, Newman TC, Sherry B, et al: Recombinant human cachectin/tumor necrosis factor but not interleukin-1a down-regulates lipoprotein lipase gene transcription in mouse 3T3-L1 adipocytes. Mol Cell Biol 1988;8:2394–2401.
35 Patton JS, Shepherd HM, Wilking H, et al: Interferons and tumour necrosis factors have similar catabolic on 3T3-L1 cells. Proc Natl Acad Sci USA 1986;83: 8313–8317.
36 Torti FM, Dieckmann B, Beutler B, et al: A macrophage factor inhibits adipocyte gene expression: an in vitro model of cachexia. Science 1985;229:867–869.
37 Feingold KR, Grunfeld C: Tumor necrosis factor-alpha stimulates hepatic lipogenesis in the rat in vivo. J Clin Invest 1987;80:184–190.
38 Warren RS, Donner DB, Starnes HF, et al: Modulation of endogenous hormone action by recombinant human tumor necrosis factor. Proc Natl Acad Sci USA 1987;84:8619–8622.
39 Tracey KJ, Wei H, Manogue KR, et al: Cachectin/TNF induces cachexia, anorexia and inflammation. J Exp Med 1988;167:1211–1227.
40 Perlmutter DH, Dinarello CA, Punsal PI, et al: Cachectin/tumor necrosis factor regulates hepatic acute phase gene expression. J Clin Invest 1986;78:1349–1353.
41 Moldawer LL, Andersson C, Gelin J, et al: Regulation of food intake and hepatic protein metabolism by recombinant-derived monokines. Am J Physiol 1988;254: 450–456.
42 Stolpen AH, Guinan EC, Fiers W: Recombinant tumor necrosis factor and immune interferon act singly and in combination to reorganize human vascular endothelial cell monolayers. Am J Pathol 1986;123:16–24.
43 Pohlman TH, Stanness KA, Beatty PG, et al: An endothelial surface factor(s) induced in vitro by lipopolysaccharide, interleukin-1, and tumor necrosis factor-α increases endothelial adherence by a CDw18-dependent mechanism. J Immunol 1986;136:4548–4553.
44 Pober JS, Bevilacqua MP, Mendrick DL, et al: Two distinct monokines, interleukin-1 and tumor necrosis factor, each independently induce biosynthesis and transient expression of the same antigen on the surface of cultured human vascular endothelial cells. J Immunol 1986;136:1680–1687.
45 Stern DM, Nawroth PP: Modulation of endothelial hemostatic properties by tumor necrosis factor. J Exp Med 1986;163:740–745.

46 Remick DG, Kunkel RG, Larrick JW, et al: Acute in vivo effects of human recombinant tumor necrosis factor. Lab Invest 1987;56:583–590.
47 Fràter-Schröder M, Risau W, Hallman R, et al: Tumour necrosis factor type alpha, a potent inhibitor of endothelial cell growth in vitro, is angiogenic in vivo. Proc Natl Acad Sci USA 1987;84:5277–5281.
48 Sugarman BJ, Aggarwal BB, Hass PE, et al: Recombinant tumor necrosis factor alpha: effects on proliferation of normal and transformed cells in vitro. Science (Wash) 1986;230:943–945.
49 Ulich TR, Castillo JD, Keys M, et al: Kinetics and mechanism of recombinant human interleukin-1 alpha and tumor necrosis factor alpha induced changes in circulating numbers of neutrophils and lymphocytes. J Immunol 1987;139:3406–3415.
50 Shalaby MR, Aggarwal BB, Rinderknecht E, et al: Activation of human polymorphonuclear neutrophil functions by interferon gamma and tumour necrosis factors. J Immunol 1985;135:2069–2073.
51 Djeu JY, Blanchard DK, Halkias D, et al: Growth inhibition of *Candida albicans* by human polymorphonuclear neutrophils: activation by interferon-γ and tumor necrosis factor. J Immunol 1986;137:2980–2984.
52 Munker R, Gassoon J, Ogawa M, et al: Recombinant human TNF induces production of granulocyte-macrophage colony stimulating factor. Nature 1986;323:729–732.
53 Philip R, Epstein LB: Tumour necrosis factor as immunomodulator and mediator of monocyte cytotoxicity induced by itself, gamma-interferon and interleukin-1. Nature 1986;323:86–89.
54 Mestan J, Digel W, Mittnacht S, et al: Antiviral effects of recombinant tumour necrosis factor in vitro. Nature 1986;323:816–819.
55 Wong GHW, Goeddel DV: Tumour necrosis factors alpha and beta inhibit virus replication and synergize with interferon. Nature 1986;323:819–821.
56 Tracey KJ, Fong Y, Hesse DG, et al: Anti-cachectin/TNF monoclonal antibodies prevent septic shock during lethal bacteremia. Nature 1987;330:662–664.
57 Bachwich PR, Chensue SW, Larrick JW, et al: Tumor necrosis factor stimulates IL-1 and PGE_2 production in resting macrophages. Biochem Biophys Res Commun 1986;136:94–101.
58 Kohase M, Henriksen-Destefano D, May LT, et al: Induction of β_2-interferon by tumor necrosis factor: a homeostatic mechanism in the control of cell proliferation. Cell 1986;45:659–666.
59 Le J, Vilcek J: Tumor necrosis factor and interleukin-1: cytokines with multiple overlapping biological activities. Lab Invest 1987;56:234–248.
60 Philip R, Epstein LB: Tumor necrosis factor as immunomodulator of monocyte cytotoxicity induced by itself, gamma-interferon, and interleukin-1. Nature 1986;323:86–89.
61 Warren RS, Donner DB, Starnes HF, et al: Modulation of endogenous hormone action by recombinant human tumor necrosis factor. Proc Natl Acad Sci USA 1987;84:8619–8622.
62 Dinarello CA: Biology of interleukin-1. FASEB J 1988;2:108–113.
63 Rothstein JL, Schreiber H: Synergy between tumor necrosis factor and bacterial products causes hemorrhagic necrosis and lethal shock in normal mice. Proc Natl Acad Sci USA 1988;85:607–611.

64 Warren RS, Starnes HF, Gabrilove JL, et al: The acute metabolic effects of tumor necrosis factor administration. Arch Surg 1987;122:1396–1400.
65 Theologides A: The anorexia-cachexia syndrome: a new hypothesis. Ann NY Acad Sci 1974;230:14–22.
66 Cerami A, Ikeda Y, Le Trang N, et al: Weight loss associated with an endotoxin-induced mediator from peritoneal macrophages: the role of cachectin (tumor necrosis factor). Immunol Lett 1985;11:173–177.
67 Michie HR, Spriggs DR, Rounds J, et al: Does cachectin cause cachexia? Surg Forum 1987;28:38–40.
68 Oliff A, Defeo-Jones D, Boyer M, et al: Tumor secreting human TNF/cachectin induce cachexia in mice. Cell 1987;50:555–563.
69 Moldawer LL, Svaninger G, Gelin J, et al: Interleukin-1 (alpha or beta) and tumor necrosis factor-alpha do not regulate protein balance in skeletal muscle. Am J Physiol 1987;253:C766–C773.
70 Starnes HF, Warren RS, Conti PS, et al: Redistribution of amino acids in rat liver and muscle induced by tumor necrosis factor requires the adrenal response. Surg Forum 1987;27:41–42.
71 Shires GT, Canizaro PC, Carrico CJ: Shock; in Schwartz SI, et al (eds): Principles of Surgery. New York, McGraw-Hill, 1979, pp 135–184.
72 Tracey KJ, Beutler B, Lowry SF, et al: Shock and tissue injury induced by recombinant human cachectin. Science (Wash) 1986;234:470–474.
73 Beutler G, Milsark IW, Cerami A: Passive immunization against cachectin/tumor necrosis factor protects mice from the lethal effect of endotoxin. Science (Wash) 1985;229:869–871.
74 Scuderi P, Lam KS, Ryan KJ, et al: Raised levels of tumor necrosis factor in parasitic infections. Lancet 1986;ii:1364–1365.
75 Waage A, Halstensen A, Espevik T: Association between tumor necrosis factor in serum and fatal outcome in patients with meningococcal disease. Lancet 1987;i:355–357.
76 Balkwill F, Osborne R, Burke F, et al: Evidence for tumor necrosis factor/cachectin production in cancer. Lancet 1987;i:1229–1232.
77 Aderka D, Fisher S, Levo Y, et al: Cachectin/tumor necrosis-factor production by cancer patients. Lancet 1985;ii:1190–1192.
78 Marano M, Fong Y, Moldawer LL, et al: Serum cachectin levels in burned and ICU patients (abstract). Am Burn Assoc, 1987.
79 Maury CPJ, Teppo A-M: Raised serum levels of cachectin/tumor necrosis factor alpha in renal allograft rejection. J Exp Med 1987;166:1132–1137.

Dr. Anthony Cerami, Laboratory of Medical Biochemistry, Rockefeller University, 1230 York Avenue, Box 277, New York, NY 10021-6399 (USA)

Sorg C (ed): Macrophage-Derived Cell Regulatory Factors.
Cytokines. Basel, Karger, 1989, vol 1, pp 89–104

Biology of the Tumor Necrosis Factors

John S. Patton, Glenn C. Rice, Gerald E. Ranges, Michael A. Palladino, Jr.

Department of Molecular Immunology, Genentech, Inc., South San Francisco, Calif., USA

Introduction

The area of cytokine biology has recently undergone a rapid expansion which has led to an understanding of many of the intricate interactions between multiple immunoregulatory proteins. As each cytokine (a term now used to denote lymphokines, growth factors, differentiation factors, suppressor factors and hormones) has become available in sufficient quantities to allow for extensive biological examinations, studies have demonstrated that these proteins possess multiple activities besides the function which led to their discovery.

Tumor necrosis factor-alpha (TNF-α) was also thought to have a narrow range of biological activities limited primarily, if not exclusively, to the induction of hemorrhagic necrosis in vivo and to direct cytotoxic actions on transformed cell lines in vitro [1]. However, since the cloning of the cDNA for TNF-α, this molecule has been shown to be a potent regulator of multiple biological systems [2] (fig. 1).

Immunoregulatory

As research on the biological activity of TNF-α progressed, it became increasingly clear that this cytokine was an important up-regulator of immune responses (fig. 2). These effects can be accomplished either by direct action of TNF-α on effector cells or by indirect action where the effect is accomplished via the induction of multiple intermediate cytokines (fig. 2).

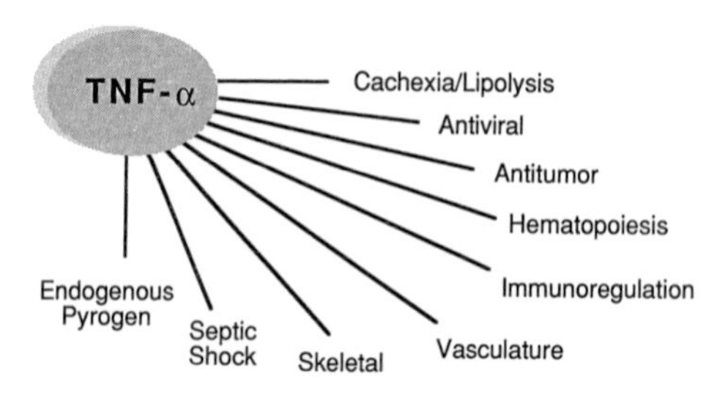

Fig. 1. Summary of biological activities regulated by TNF-α.

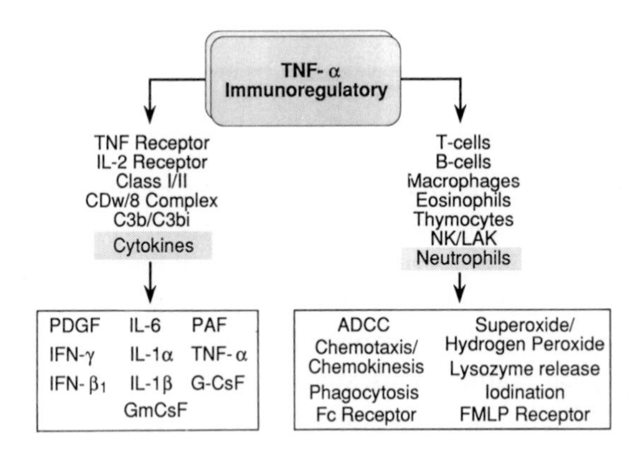

Fig. 2. Summary of immunoregulatory activities regulated by TNF-α.

The effects of TNF-α either on immune responses or tumors are often greatly amplified in the presence of other cytokines such as interferon-gamma (IFN-γ) [3, 4].

Stimulation of endothelial cells with TNF-α results in the production of interleukin-1 (IL-1) and the enhanced expression of class I and II histocompatibility antigens [5, 6]. In addition, endothelial cells exposed to TNF-α express antigens which enhance adherence to neutrophils [7] and lymphocytes to these cells [8]. Such activities may direct immunologically relevant cells to the antigenic site and induce the local release of substances which directly regulate the immune response.

One of the first immunoactivities defined for TNF-α was the ability to enhance numerous functions of polymorphonuclear neutrophils (PMN) [9–12] (fig. 2). Recently, Silberstein and David [13] have shown that TNF-α also enhances the cytotoxic activity of eosinophils. Macrophages, the major cell source of TNF-α, are also remarkably responsive to the stimulatory effects of TNF-α. Talmadge et al. [14] and Hori et al. [15] have shown enhanced cytolytic function of activated macrophages after either in vitro or in vivo exposure to TNF-α. Bachwich et al. [16] have shown that, in vitro, TNF-α can stimulate monocytes to produce prostaglandin E2 and IL-1. In further studies, Philip and Epstein [17] reported that TNF-α induces monocytes to produce IL-1 and, in an autocrine fashion, more TNF-α. Recently, Scheurich et al. [18] have shown that TNF-α can enhance IFN-γ production, however, only in previously stimulated mononuclear cell cultures. All three of these cytokines can enhance monocyte cytotoxicity against selected tumor targets and this toxicity can be blocked by anti-TNF-α antibodies [17]. Thus, TNF-α can remarkably induce its own production while also mediating an effector function of the macrophage.

Ostensen et al. [19] have reported that the activity of human natural killer (NK) cells can be augmented by preincubation with TNF-α and can be enhanced even further if interleukin-2 (IL-2) is added to the cultures. In these experiments, the target cells were not directly susceptible to TNF-α cytotoxicity as in the case with the macrophage. These investigators reported that IL-2 receptor levels on the NK cells were increased by TNF-α exposure. However, this observation does not fully explain the effects of TNF-α on NK activity since augmentation also occurs in the absence of IL-2. In addition, these investigators showed that the enhanced NK activity was not due to TNF-α induction of IFN-γ [19]. Ortaldo et al. [20] have shown that the cytotoxic activity of natural cytotoxic cells, which differ from NK cells by cell surface markers and the pattern of susceptible targets, is also mediated by TNF-α.

Recent studies have shown that TNF-α can synergize with IL-2 to enhance the induction of human lymphokine-activated killer (LAK) cell activity [21, 22]. Although IL-2 is the requisite cytokine in this system, Owen-Schaub et al. [21] have shown, using TNF-α specific antisera, that endogenous TNF-α is produced during the course of IL-2 activation of LAK precursors. Thus, the exogenous TNF-α apparently augments the normal process of IL-2 driven activation rather than altering it. One effect of TNF-α on this system is the increase in IL-2 receptor expression on the effector cells. Curiously, this results in the enhancement of LAK effector activity without a concomitant enhancement in their proliferative response [21]. Recently, Espevik et al. [22] have indicated that transforming growth factors (TGF)-β_1 and -β_2 are capable of suppressing IL-2 activation of LAK cells and that this suppression was significantly reversed by the addition of TNF-α to the cultures. As in the previous investigation [21], these researchers reported that IL-2 induces TNF-α production in LAK cell cultures and that the production of TNF-α was suppressed in the presence of TGF-β_1 [22]. Taken together, these results also point to a critical role for TNF-α in LAK cell development.

Reports concerned with the effects of TNF-α on B lymphocyte function have been rare. Talmadge et al. [14] demonstrated that in vivo treatment with TNF-α can enhance antibody production to the T cell-dependent antigen, bovine serum albumin. In a series of well-designed experiments, Kehrl et al. [23] and Jelinek and Lipsky [24] have shown that TNF-α augments the specific proliferative response of purified human tonsillar-derived B cells to *Staphylococcus aureus* Cowan and enhances IL-2 signaled IgM production. However, in a study of pokeweed mitogen-induced B cell differentiation, Kashiwa et al. [25] demonstrated a suppressive effect of TNF-α on immunoglobulin synthesis. The major technical difference between these studies [23–25] is the presence of T cells in the latter study. Thus, the observed suppression may have been due to the induction of T suppressor cells and/or TGF-β which can also suppress B cell functions [26, 27].

The ability of T cells to produce TNF-α and the concomitant effects of TNF-α on T cell function have only recently come under study. Cuturi et al. [28] have demonstrated that both CD4 and CD8 T cells activated with either mitogens or phorbol diester and calcium ionophore produce TNF-α independent of any monocyte contamination. In a recent report, Scheurich et al. [18] demonstrated that TNF-α enhanced both IL-2-dependent T cell proliferation and IFN-γ production by activated mononuclear cells. The increased proliferation was ascribed to amplified IL-2 receptor expression induced by TNF-α. Furthermore Yokota et al. [29] have described direct enhancement by TNF-α on both the CD4 and CD8 cell proliferative response to mitogen or antigen. These observed effects were explained by an increase in IL-2 receptor expression. In both studies the effects of TNF-α were only observed on activated T cells, supporting previous studies which demonstrated that TNF-β failed to stimulate IFN-γ production in nonactivated T cells [18, 30]. Ranges et al. [31] have published that TNF-α can act as a comitogen with PHA-P in the induction of murine thymocyte proliferation. Although TNF-α was unable to stimulate cell division in the absence of either IL-1 or IL-2, it was able to do so in their presence without the comitogenic effect of PHA-P.

There are two reports which indicate that TNF-α can augment T cell proliferation and cytolytic T cell generation in a mixed lymphocyte reaction (MLR). Ranges et al. [32] have reported that significant levels of TNF-α are generated during the early stages of a murine MLR and that polyclonal antibody to murine TNF-α will suppress the development of cytolytic T cells when added to the MLR cultures. It was also demonstrated that TNF-α significantly enhanced cytolytic T cell activity in MLR cultures established under suboptimal conditions. In the same report [32], TGF-β was shown to suppress cytolytic T cell generation and that TNF-α was able to partially reverse this suppression. Shalaby et al. [33] have now shown similar enhancing effects of both TNF-α and TNF-β on the human MLR and have shown that TNF-β also enhances IL-2 receptor expression of T cells [33]. As in the murine MLR studies, antibodies to human TNF-α (but not TNF-β) significantly suppressed the proliferative response in human MLR [33]. Therefore, the mechanism of TNF-α/TNF-β-mediated enhancement of T cell responses may involve the enhancement of IL-2 receptors, class I and II expression [6, 18, 34–36] and/or cytokine production (fig. 2) such as IL-1, IFN-γ and IL-6 [5, 18, 37].

The importance of TNF-α to allogenic reactions has been underscored by two reports which implicate TNF-α during in vivo allograft rejection and acute graft-versus-host disease (GVHD). Maury and Teppo [38] monitored circulating TNF-α levels in patients undergoing renal transplant surgery. A strong correlation of high TNF-α levels and acute graft rejection episodes was observed, the highest levels being associated with irreversible rejections. In an elegant histological analysis of murine GVHD, Piguet et al. [39] demonstrated a central role for TNF-α by demonstrating the presence of TNF-α in skin and intestinal lesions during GVHD and the prevention of such lesions by anti-TNF-α pretreatment.

The numerous biological effects of TNF-α are mediated through high affinity receptors which are present on both normal and transformed cell lines. In most cell lines examined, receptor number varies between 200 and 20,000 per cell [40–45]. The presence of receptors for TNF-α, while being necessary, is not sufficient to make cells susceptible to its actions. Both TNF-α and TNF-β apparently share a common specific receptor and binding can be correlated with cell lysis from either molecule [42]. Cross-linking studies on several human tumor cell lines using disuccinimidyl suberate have shown binding of two polypeptides, 90 and 75 kd to ^{125}I-TNF [44, 46]. However, Creasey et al. [40], using 1,5-difluoro-2,4-dinitrobenzene to affinity label the TNF-α receptor, detected two additional polypeptides, of 138 and 54 kd. Most striking was their finding that the 138-kd polypeptide was observed only in cells sensitive to the cytotoxic activity of TNF-α such as the human breast carcinoma cells MCF-7 and was absent in variants resistant to the effects of TNF-α. Taken together, these data suggest that factors other than receptor number are necessary for the cytotoxic activity of TNF-α.

Pretreatment of several human tumor cell lines with IFN-γ can increase specific binding of TNF-α to its receptors without a change in affinity constant [41, 42, 47]. This effect was originally thought to contribute to the potent cytotoxic synergy between IFN-γ and TNF-α seen in some tumor cell lines with low receptor numbers. However, data from Tsujimoto et al.[48] suggest that TNF-α receptor up-regulation by IFN-γ is not a major mechanism of IFN-γ and TNF-α cytotoxic synergy. These authors found that IFN-β decreased the IFN-γ-mediated TNF-α receptor increase, but did not inhibit the cytotoxic enhancement of TNF-α and IFN-γ in the human colon carcinoma HT-29 cell line. Furthermore, Ruggerio et al. [47] showed that SK-MEL-109 cell growth was inhibited synergistically with TNF-α and IFN-γ, even though IFN-γ did not increase TNF-α receptor numbers. Thus, the IFN-γ-mediated increase in TNF-α receptors is not necessarily associated with a concomitant increase in TNF-α cytotoxicity. In contrast to the effects of IFN-γ, IL-1, a cytokine whose production can also be regulated by TNF-α [3], down-regulates the expression of TNF-α receptors [41, 49]. It would be of interest to examine the molecular structure of the TNF-α receptor after up-regulation with IFN-γ, specifically the possible changes in the association of the 138-kd polypeptide as detected by Creasey et al. [40] and its possible role in modulating cytotoxic interaction.

We have developed new methodologies to detect TNF receptors on mammalian cells using fluorescence-activated cell sorting (FACS) which may lead to novel approaches to examine some of these issues. TNF receptors were detected by either (a) direct method of treating cells with biotinylated TNF-α, or (b) by an indirect method of first labeling cells with TNF-α, followed by binding with antibodies against TNF-α. Both methods permit the sensitive detection of approximately 1,000 TNF-α receptors per cell (fig. 3). Scatchard analysis with [^{125}I]-TNF-α revealed that the high sort U937 variants expressed 5-fold more receptors per cell than the parental cell line (from 10,000 to 50,000), without a change in binding affinity (4×10^{-10} *M*). As with the parent, TNF-β will compete with TNF-α for receptor binding (fig. 3).

'Cachexia'

Beutler et al. [50] originally described a factor from activated macrophages, cachectin, based on its in vitro activity against adipocytes in culture.

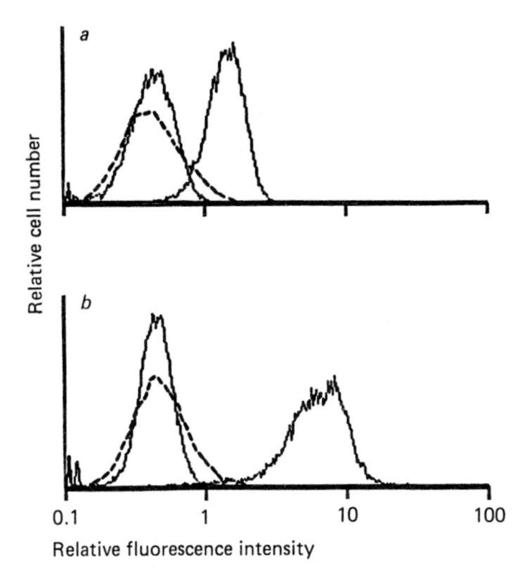

Fig. 3. FACS distributions for TNF-α receptors on U937 cells. Cells were stained using the indirect approach described in the text. *a* Distributions for parental U937 cells. The solid curve on the left is for cells stained without the addition of TNF-α, the solid curve on the right with TNF-α. The dashed curve represents staining with TNF-α in the presence of a 1,000-fold excess of TNF-β. The lack of staining is due to the fact that TNF-α and TNF-β apparently share a common receptor [42]. *b* Fluorescence distribution of U937 cells that were sorted 20 times for the upper 1% fluorescence for TNF-α receptors. The curve on the left, without TNF-α; that on the right with the addition of TNF-α. The dashed line represents cells stained with TNF-α but in the presence of 1,000-fold excess of TNF-β.

This factor was subsequently shown to be TNF-α [50]. Cachectin inhibits the expression of a family of adipocyte genes causing reduction in the levels of lipoprotein lipase (LPL), the enzyme responsible for bringing fatty acids into the adipocyte, and the enzymes responsible for de novo fat synthesis [51]. These studies suggested that since LPL levels are low and serum triglycerides are elevated in infection [52], TNF was a possible mediator of wasting (cachexia).

Acute Toxic Effects

Within 4 h after the administration of high doses of rHuTNF-α (250 μg/kg i.v.) to rats, severe gastrointestinal inflammation with associated cecal hemorrhage and blockage of gastric emptying was evident [53]. Although weight loss associated with these acute effects occurred in the first 2 days, with continued daily injections tolerance developed and cachexia did not occur. If lower, nonhemorrhagic doses of TNF-α were administered, weight loss was minimal. The inflammatory effects of TNF-α in the gastrointestinal tract are similar to those induced by recombinant TNF-α in Meth A tumors [54] and in other organs following systemic administration of high doses of natural TNF-α [55]. It is not known why the bowel and the new vasculature of the implanted Meth A tumor are especially sensitive to TNF-α. The initial lesions appear to be vascular since Meth A tumor cells are resistant to direct TNF-α-induced cytotoxicity [54]. In the gastrointestinal tract, TNF-α may stimulate the production of platelet-activating factor (PAF), an endogenous mediator of bowel necrosis in endotoxemia (fig. 2) [56]. In addition, since endotoxin increases the toxicity of TNF-α [57], the gut may be particularly sensitive because of the high levels of endotoxin normally present there.

Acute Effects on Lipid Metabolism

Infection and tumorigenesis are often accompanied by high serum triglyceride levels, increased liver lipogenesis and decreased synthesis and uptake of fat by adipose tissue [52]. There is good evidence that rHuTNF-α can acutely induce most if not all of these effects. Feingold and Grunfeld [58] have shown that 2h after administration of rHuTNF-α to rats (125 μg/kg), plasma triglycerides increased 2.2-fold and remained elevated for at least 17 h. rHuTNF-α rapidly stimulated incorporation of tritiated water into fatty acids in the liver (1–2 h), which persisted for 17 h. In addition, TNF-α rapidly increased the concentration of labeled fatty acids in the plasma, raising the possibility that stimulation of hepatic lipogenesis by TNF-α contributes to the hyperlipidemia of infection [58]. In studies with streptozotocin-induced diabetic rats in which LPL levels are markedly suppressed, Grunfeld et al. [59] found that TNF-α elevated serum triglycerides and stimulated liver lipogenesis without affecting LPL. This suggests that the elevation in serum lipid seen with TNF-α can be explained by increased liver synthesis alone. Semb et al. [60] have shown that a single intravenous dose of murine TNF-α caused a suppression of LPL levels in adipose tissue of fed rats, mice and guinea pigs for 48 h. While TNF-α decreased LPL activity only in adipose tissue, it did increase the overall postheparin LPL activity in the heart, lung and most markedly in the liver [60]. Thus some, if not most of the alterations in lipid metabolism that are seen in infection can be, at least acutely, induced by TNF-α.

Other Cytokines Possess 'Cachectic' Properties

The ability of rHuTNF-α to inhibit lipid anabolism in adipocytes is a general property shared with other cytokines including TNF-β, IFN-α, IFN-β, IFN-γ and IL-1 [61]. The net effect of cytokine action on fat cells is catabolic; fat uptake and synthesis are inhibited and fat mobilization is stimulated. TNFs and IFNs appear to act on adipocytes through separate receptors [61]. Reports that TNF-α does not effect lipogenesis in adipocytes [62] may be explained by differences in the 3T3-L1 subclones utilized [C. Grunfeld, pers. commun.].

Sensitization, Tolerance and Escalating Doses

Some growing tumors can induce marked hypersensitivity to TNF-α and endotoxin [63]. While the administration of 500 μg of rHuTNF-α was lethal to normal mice, 4 and 0.01 μg, respectively, on day 15 following tumor inoculation was lethal in 50% of the animals bearing EMT6 or Lewis lung carcinoma tumors [63]. Sensitization was paralleled by marked granulocytosis. Tumor-bearing animals often possess an activated reticuloendothelial system [64] that has induced a preparative state for shock and the generalized Shwartzman reaction. The accumulation of activated macrophages and granulocytes in cancerous animals would lead, after TNF-α challenge, to the release of active substances (including additional TNF-α) (fig. 3) which could mediate systemic toxicity [63]. The exact mechanisms involved in sensitization are still unknown.

The development of tolerance (tachyphylaxis) to the many effects of TNF-α is similar to the tolerance seen in humans to some of the adverse side effects of alpha interferon (IFN-α). Sherwin et al. [65] reported tachyphylaxis with fever, flu-like symptoms and gastrointestinal effects, whereas fatigue and anorexia tended to be cumulative. Lau et al. [66] have shown in vivo and in vitro that IFN-α receptor number on human blood lymphocytes decreased nearly 10-fold after interferon treatment and required roughly 72 h to recover. During the recovery period the cells also exhibited hyporesponsiveness to interferon treatment [65]. Thus, down-regulation of TNF-α receptors is a possible cause of the tolerance seen in this study as dosing of rats with TNF-α at 5- to 10-day intervals (250 μg/kg i.v.) allowed sensitivity to toxicity to be maintained [53].

Nude mice injected intramuscularly with CHO cells transfected with the gene for TNF-α developed severe cachexia and weight loss beginning at about 20 days post-inoculation [67] while mice implanted with parental CHO cells maintained their weight. This study therefore provides evidence that TNF-α can be cachectic when secreted by a tumor. Moreover, as it must be assumed that the TNF-α animals received a constantly escalating dose of the cytokine as the tumor grew, tachyphylaxis may not have been possible. It is already known that tumors can hypersensitize animals to TNF-α, thus some additional factors or conditions may have prevailed in the TNF tumor group to cooperate with TNF-α to cause wasting. Clearly there is more to cachexia than TNF-α alone as other cytokines possess similar lipolytic effects [61] and repeated injections of constant doses of TNF-α is not cachectic in healthy mice. Future efforts should be focused on lipolytic factors released by tumor cells and cofactors which block the development of tolerance to TNF-α.

Antitumor Activities

It is now clear that multiple processes can be involved in the initiation of tumor necrosis and the resulting specific rejection of tumor implants (fig. 2). This fact becomes clearer when one considers that transformed cell lines which are not growth inhibited by TNF-α in vitro can be exquisitely sensitive in vivo [54, 68, 69]. The in vivo antitumor actions of TNF-α are now known to involve the activation of multiple host cells including PMNs, endothelial cells and lymphocytes. The first stage hemorrhagic necrosis apparently involves the activation of neutrophils and the induction of changes in the homeostatic properties of endothelial cells and capillaries in the tumor leading to PMN adherence and thrombosis formation [5–7, 9–12, 54].

The induction of hemorrhagic necrosis in sarcoma-bearing mice appears to require the participation of T cells as nude mice or T cell-depleted mice show reduced hemorrhagic tumor necrosis and antitumor effects after TNF-α treatment [68, Figari et al., manuscript submitted]. Similar results have recently been published using endotoxin as the antitumor agent [70], further supporting the relationships between the mechanisms of action for endotoxin and TNF-α [1, 70]. In addition, studies by North and Bursuker [71] have demonstrated that during the first 7 days of Meth A growth, Lyt-2-positive T cells were responsible for the development of concomitant immunity, supporting our data that cytotoxic T cells to Meth A can be generated in mice within 7 days of immunization [72]. However, this response is apparently not sufficient to induce complete rejection after TNF-α treatment as an L3T4-mediated response presumably similar to that of a delayed hypersensitivity response to protein antigens is responsible for the ultimate rejection of Meth A [Figari et al., manuscript submitted].

We feel this tumor model is appropriate to work out the multiple stages of TNF-α-induced tumor regression as it takes into account the immunoregulatory effects of TNF-α. Although nude mice tumor xenograft models yield useful information about the direct antitumor effects of TNF-α [73], the effects we have outlined in this review on the immune system are not possible to evaluate in these models.

The recently defined species specificity of murine versus human TNF-α on murine tumor cells [74, 75] and thymus cells [31] must now be considered when characterizing the biological activities of TNF-α in vitro and in vivo. It is tempting to speculate that in the murine tumor models where human TNF-α is less active than murine TNF-α, that these differences are in part due to the failure of the human TNF-α to invoke the T cell arm of the immune system.

Conclusion

As with many of the cytokines for which new information is accumulating quickly, the capabilities of TNF-α and TNF-β have far surpassed the function for which the molecule was originally named. The initially described antitumor effects of these molecules are now thought to be the result of a constellation of activities directed at several cell types involved in tumor rejection. As is also apparent, the systemic distribution of TNF-α in high concentrations can result in deleterious effects on the homeostatic balance in vivo and thus complicates its use in a clinical setting. However, as we learn more about the many important biological activities of these two molecules, we will discover how they may be most effectively utilized in a variety of therapeutic functions. Meanwhile, as new activities of TNF-α and TNF-β are discovered, it is becoming clear that 'Tumor necrosis factors are more than tumor necrosis factors'.

References

1 Williamson B, Carswell E, Rubin B, et al: Human tumor necrosis factor produced by human B-cell lines: Synergistic cytotoxin interaction with human interferon. Proc Natl Acad Sci USA 1983;80:5397–5401.
2 Pennica D, Nedwin GE, Hayflick JS, et al: Human tumor necrosis factor: precursor structure, cDNA cloning, expression, and homology to lymphotoxin. Nature 1984;321:724–729.
3 Dinarello CA, Cannon JG, Wolff SM, et al: Tumor necrosis factor (cachectin) is an endogenous pyrogen and induces production of interleukin-1. J Exp Med 1986; 163:1433–1450.
4 Sugarman BJ, Aggarwal BB, Hass PE, et al: Recombinant human tumor necrosis factor-alpha effects on proliferation of normal and transformed cells in vitro. Science 1985;230:943–945.
5 Nawroth P, Bank I, Handley D, et al: Tumor necrosis factor/cachectin interacts with endothelial cell receptors to induce release of interleukin-1. J Exp Med 1983; 163:1363–1375.
6 Collins T, Lapierre L, Fiers W, et al: Recombinant human tumor necrosis factor increases mRNA levels and surface expression of HLA-A,B antigens in vascular endothelial cells and dermal fibroblasts in vitro. Proc Natl Acad Sci USA 1986; 83:446–450.
7 Gamble J, Harlan J, Klebanoff S, et al: Stimulation of the adherence of neutrophils to umbilical vein endothelium by human recombinant tumor necrosis factor. Proc Natl Acad Sci USA 1985;82:8667–8671.
8 Pober J, Gimbrone M, Lapierre L, et al: Overlapping patterns of activation of human endothelial cells by interleukin-1, tumor necrosis factor and immune interferon. J Immunol 1986;137:1893–1896.
9 Figari I, Mori N, Palladino M Jr: Regulation of neutrophil chemotaxis and superoxide production by recombinant tumor necrosis factor-alpha and beta comparison to recombinant interferon-γ and interleukin-1α. Blood 1987;70:979–984.
10 Shalaby M, Aggarwal B, Rinderknecht E, et al: Activation of human polymorphonuclear neutrophil functions by interferon-γ and tumor necrosis factors. J Immunol 1985;135:2069–2073.
11 Klebanoff S, Vadas MA, Harlan JM, et al: Stimulation of neutrophils by tumor necrosis factor. J Immunol 1986;136:4220–4225.
12 Nathan CF: Neutrophil activation on biological surfaces: Massive secretion of hydrogen peroxide in response to product of macrophages and lymphocytes. J Clin Invest 1987;80:1550–1560.
13 Silberstein D, David J: Tumor necrosis factor enhances eosinophil toxicity to *Schistosoma mansoni* larvae. Proc Natl Acad Sci USA 1986;83:1055–1059.
14 Talmadge J, Phillips H, Schneider M, et al: Immunomodulatory properties of recombinant murine and human tumor necrosis factor. Cancer Res 1988;48:544–550.
15 Hori K, Ehrke MJ,, Mace K, et al: Effect of recombinant human tumor necrosis factor on the induction of murine macrophage tumoricidal activity. Cancer Res 1987;47:2793–2798.
16 Bachwich P, Chensue S, Larrick J, et al: Tumor necrosis factor stimulates interleukin-1 and prostaglandin E_2 production in resting macrophages. Biochem Biophys Res Commun 1986;136:94–101.

17 Philip R, Epstein L: Tumor necrosis factor as immunomodulator and mediator of monocyte cytotoxicity induced by itself, γ-interferon and interleukin-1. Nature 1986;323:86–89.
18 Scheurich P, Thoma B, Ucer U, et al: Immunoregulatory activity of recombinant human tumor necrosis factor (TNF)-α: induction of TNF receptors on human T cells and TNF-α mediated enhancement of T cell responses. J Immunol 1987;138:1786–1790.
19 Ostensen M, Thiele D, Lipsky P: Tumor necrosis factor-α enhances cytolytic activity of human natural killer cells. J Immunol 1987;138:4185–4191.
20 Ortaldo J, Mason L, Mathieson B, et al: Mediation of mouse natural cytotoxic activity by tumor necrosis factor. Nature 1986;321:700–702.
21 Owen-Schaub L, Gutterman J, Grimm E: Synergy of tumor necrosis factor and interleukin-2 in the activation of human cytotoxic lymphocytes: effect of tumor necrosis factor-α and interleukin-2 in the generation of human lymphokine-activated killer cell cytotoxicity. Cancer Res 1988;48:788–792.
22 Espevik T, Figari I, Ranges GE, et al: Transforming growth factor-β_1 (TGF-β_1) and recombinant human tumor necrosis factor-α reciprocally regulate the generation of lymphokine-activated killer cell activity. Comparison between natural porcine platelet-derived TGF-β_1 and TGF-β_2, and recombinant human TGF-β_1. J Immunol 1988;140:2312–2316.
23 Kehrl J, Miller A, Fauci A: Effect of tumor necrosis factor-α on mitogen-activated human B cells. J Exp Med 1987;166:786–791.
24 Jelinek DF, Lipsky PE: Enhancement of human B cell proliferation and differentiation by tumor necrosis factor-α and interleukin-1. J Immunol 1987;139:2970–2976.
25 Kashiwa H, Wright S, Bonavida B: Regulation of B cell maturation and differentiation. I. Suppression of pokeweed mitogen-induced B cell differentiation by tumor necrosis factor (TNF). J Immunol 1987;138:1383–1390.
26 Lee G, Ellingsworth LR, Gi S, et al: β-Transforming growth factors are potential regulators of B lymphopoiesis. J Exp Med 1987;166:1290–1299.
27 Kehrl JH, Roberts AB, Wakefield LM, et al: Transforming growth factor β is an important immunomodulatory protein for human B lymphocytes. J Immunol 1986;137:3855–3860.
28 Cuturi M, Murphy M, Costa-Giomi M, et al: Independent regulation of tumor necrosis factor and lymphotoxin production by human peripheral blood lymphocytes. J Exp Med 1987;165:1581–1594.
29 Yokota S, Geppert T, Lipsky P: Enhancement of antigen- and mitogen-induced human T lymphocyte proliferation by tumor necrosis factor-α. J Immunol 1988;140:531–536.
30 Svedersky LP, Nedwin GE, Goeddel DV, et al: Interferon-γ enhances induction of lymphotoxin in recombinant interleukin-2-stimulated peripheral blood mononuclear cells. J Immunol 1985;134:1604–1608.
31 Ranges GE, Zlotnik A, Espevik T, et al: Tumor necrosis factor α/cachectin is a growth factor for thymocytes. Synergistic interactions with other cytokines. J Exp Med 1988;167:1472–1478.
32 Ranges GE, Figari I, Espevik T, et al: Inhibition of cytotoxic T cell development by transforming growth factor β and reversal by recombinant tumor necrosis factor α. J Exp Med 1987;166:991–998.

33 Shalaby MR, Espevik T, Rice GC, et al: The involvement of human tumor necrosis factors alpha and beta in the mixed lymphocyte reaction. J Immunol 1988;141:499–503.
34 Klyczek KK, Murasko DM, Blank KJ: Interferon-γ, interferon-α/β, and tumor necrosis factor differentially affect major histocompatibility complex class I expression in murine leukemia virus-induced tumor cell lines. J Immunol 1987;139:2641–2648.
35 Chang RJ, Lee SH: Effects of interferon-γ and tumor necrosis factor-α on the expression of an Ia antigen on a murine macrophage cell line. J Immunol 1986;137:2853–2856.
36 Pfitzenmaier K, Scheurich P, Schlüter-Krönke M: Tumor necrosis factor enhances HLA-A,B,C and HLA-DR gene expression in human tumor cells. J Immunol 1987;138:975–980.
37 Kohase M, Henriksen-DeStefano D, May LT, et al: Induction of β_2-interferon by tumor necrosis factor: a homeostatic mechanism in the control of cell proliferation. Cell 1986;45:659–666.
38 Maury CPJ. Teppo AM: Raised serum levels of cachectin/tumor necrosis factor-α in renal allograft rejection. J Exp Med 1987;166:1132–1137.
39 Piguet P, Grau G, Allet B, et al: Tumor necrosis factor/cachectin is an effector of skin and gut lesions of the acute phase of graft-vs-host disease. J Exp Med 1987;166:1280–1289.
40 Creasey A, Yamamoto R, Vitt C: A high molecular weight component of the human tumor necrosis factor receptor is associated with cytotoxicity. Proc Natl Acad Sci USA 1987;84:3293–3297.
41 Tsujimoto M, Yip Y, Vilcek J: Interferon-γ enhances expression of cellular receptors for tumor necrosis factor. J Immunol 1986;136:2441–2444.
42 Aggarwal B, Eessalu T, Hass P: Characterization of receptors for human tumour necrosis factor and their regulation by γ-interferon. Nature 1985;318:665–667.
43 Rubin B, Anderson S, Sullivan S, et al: High affinity binding of ^{125}I-labeled human tumor necrosis factor (LukII) to specific cell surface receptors. J Exp Med 1985;162:1099–1104.
44 Kull F, Jacobs S, Cuatrecasas P: Cellular receptor for ^{125}I-labeled tumor necrosis factor: specific binding, affinity labeling, and relationship to sensitivity. Proc Natl Acad Sci USA 1985;82:5756–5760.
45 Baglioni C, McCandless S, Tavernier J, et al: Binding of human tumor necrosis factor to high affinity receptors on HeLa and lymphoblastoid cells sensitive to growth inhibition. J Biol Chem 1985;260:13395–13397.
46 Israel S, Hahn T, Holtmann H, et al: Binding of human TNF-α to high-affinity cell surface receptors: effect of IFN. Immunol Lett 1986;12:217–224.
47 Ruggerio V, Tavernier J, Fiers W, et al: Induction of the synthesis of tumor necrosis factor receptors by interferon-γ. J Immunol 1986;136:2445–2450.
48 Tsujimoto M, Feinman R, Vilcek J: Differential effects of type I IFN and IFN-γ on the binding of tumor necrosis factor to receptors in two human cell lines. J Immunol 1986;137:2272–2276.
49 Holtmann H, Wallach D: Down-regulation of the receptors for tumor necrosis factor by interleukin-1 and 4β-phorbol-12-myristate-13-acetate. J Immunol 1987;139:1161–1167.
50 Beutler B, Greenwald D, Hulmes JD, et al: Identity of tumor necrosis factor and the macrophage-secreted factor cachectin. Nature 1985;316:552–554.

51 Torti FM, Dieckmann B, Beutler B, et al: A macrophage factor inhibits adipocyte gene expression: An in vitro model for cachexia. Science 1985;229:867–871.
52 Bisel WR: Metabolic response to infection. Ann Rev Med 1975;26:9–20.
53 Patton JS, Peters PM, McCabe J, et al: Development of partial tolerance to the gastrointestinal effects of high doses of recombinant tumor necrosis factor-α in rodents. J Clin Invest 1987;80:1587–1596.
54 Palladino MA, Shalaby MR Kramer SM, et al: Characterization of the antitumor activities of human tumor necrosis factor-alpha and the comparison with other cytokines: Induction of tumor specific immunity. J Immunol 1987;138:4023–4032.
55 Tracey KJ, Beutler B, Lowry SF, et al: Shock and tissue injury induced by recombinant human cachectin. Science 1986;234:470–474.
56 Hsueh W, Gonzalez-Crussi F, Arroyave JL: Platelet-activating factor: an endogenous mediator for bowel necrosis in endotoxemia. FASEB J 1987;1:403–405.
57 Rothstein JL, Schreiber H: Synergy between tumor necrosis factor and bacterial products causes hemorrhagic necrosis and lethal shock in normal mice. Proc Natl Acad Sci USA 1988;85:607–611.
58 Feingold KR, Grunfeld C: Tumor necrosis factor-α stimulates hepatic lipogenesis in the rat in vivo. J Clin Invest 1987;80:184–190.
59 Feingold KR, Soued M, Staprans I, et al: The effect of TNF on lipid material in the diabetic rat: Evidence that inhibition of adipose tissue lipoprotein lipase activity is not required for TNF induced hyperlipidemia. J Clin Invest, in press.
60 Semb H, Peterson J, Tavernier J, et al: Multiple effects of tumor necrosis factor on lipoprotein lipase in vivo. J Biol Chem 1987;262:8390–8394.
61 Patton JS, Shepard HM, Wilking H, et al: Interferons and tumor necrosis factors have similar catabolic effects on 3T3L1 cells. Proc Natl Acad Sci USA 1986; 83:8313–8317.
62 Price SR. Olivecrona T, Pekala PH: Regulation of lipoprotein lipase synthesis by recombinant tumor necrosis factor: The primary regulatory role of the hormone in 3T3L1 adipocytes. Arch Biochem Biophys 1986;251:738–746.
63 Bartholeyns J, Freuenberg M, Galanos C: Growing tumors induce hypersensitivity to endotoxin and tumor necrosis factor. Infect Immun 1987;55:2230–2233.
64 Bases R, Krakoff I: Enhanced reticuloendothelial phagocytic activity in myeloproliferative diseases. J Reticuloendothel Soc 1965;2:1–7.
65 Sherwin SA, Knost JA, Fein S, et al: A multiple-dose phase I trial of recombinant leukocyte A interferon in cancer patients. J Am Med Assoc 1982;248:2461–2466.
66 Lau AS, Hannigan GE, Freedman MH, et al: Regulation of interferon receptor expression in human blood lymphocytes in vitro and during interferon therapy. J Clin Invest 1986;77:1632–1638.
67 Oliff A, Defoe-Jones D, Boyer M, et al: Tumors secreting human TNF/cachectin induce cachexia in mice. Cell 1987;50:555–563.
68 Haranaka K, Satomi N, Sakurai A: Antitumor activity of murine tumor necrosis factor (TNF) against transplanted murine tumors and heterotransplanted human tumors in nude mice. Int J Cancer 1984;34:263–267.
69 Creasey AA, Reynolds MT, Laird W: Cures and partial regression of murine and human tumors by recombinant human tumor necrosis factor. Cancer Res 1986;46:5687–5690.

70 North RJ, Havell EA: The antitumor function of tumor necrosis factor (TNF). II. Analysis of the role of endogenous TNF in endotoxin-induced hemorrhagic necrosis and repression of an established sarcoma. J Exp Med 1988;167:1086–1099.
71 North RJ, Bursuker I: Generation and decay of the immune response to a progressive fibrosarcoma. J Exp Med 1984;159:1295–1311.
72 Palladino MA, Srivastava PK, Oettgen HF, et al: Expression of a shared tumor-specific antigen by two chemically induced BALB/c sarcomas. Cancer Res 1987;47:5074–5079.
73 Balkwill FR, Lee A, Aldam G, et al: Human tumor xenografts treated with recombinant human tumor necrosis factor alone or in combination with interferons. Cancer Res 1986;46:3990–3993.
74 Brouckaert PGG, Leroux-Roels GG, Guisez Y, et al: In vivo anti-tumor activity of recombinant human and murine TNF, alone and in combination with murine IFN-γ, on a syngeneic murine melanoma. Int J Cancer 1986;38:763–769.
75 Kramer SM, Aggarwal BB, Eessalu TE, et al: Characterization of the in vitro and in vivo species preference of human and murine tumor necrosis factor-α. Cancer Res 1988;48:920–925.

Michael A. Palladino, Jr., Department of Molecular Immunology, Genentech Inc., 460 Point San Bruno Boulevard, South San Francisco, CA 94080 (USA)

Sorg C (ed): Macrophage-Derived Cell Regulatory Factors.
Cytokines. Basel, Karger, 1989, vol 1, pp 105–154

Interleukin-1 and Its Related Cytokines

Charles A. Dinarello

Division of Geographic Medicine and Infectious Diseases, Boston, Mass., USA

Introduction

Interleukin-1 (IL-1) is the term for two polypeptides (IL-1α and IL-1β) that possess a wide spectrum of immunologic and nonimmunologic activities. Although both forms of IL-1 are distinct gene products, they recognize the same receptor and share the same biological properties. IL-1 is produced in response to infection, microbial toxins, inflammatory agents, products of activated lymphocytes, complement, and clotting components. Its name as an interleukin, which means 'between' leukocytes, is somewhat inappropriate because IL-1 is synthesized by both leukocytic as well as nonleukocytic cells; furthermore, IL-1 effects are not restricted to leukocytes but rather are manifested in nearly every tissue. Nevertheless, the term 'interleukin' is often used for lack of a better system of nomenclature. To date, the primary amino acid sequences for 8 molecules of human origin have been reported (IL-1α, IL-1β, IL-2 thru IL-7); most of these molecules share some biological activities.

There are other polypeptides for which the primary human amino acid sequences are known but are not called interleukins. IL-1 is the prototype of a group of biologically potent polypeptides with molecular weights between 10 and 30 kd. Because these substances are produced by a variety of cells and act on many different cell types, there is a growing acceptance to call these polypeptides 'cytokines' rather than 'lymphokines' or 'monokines'. Of the various 'cytokines', several share the ability to stimulate or augment cell proliferation, initiate the synthesis of new proteins in a variety of cells, and induce the production of inflammatory metabolites. Although several exist, IL-1 is biologically similar to tumor necrosis factor (TNF), lymphotoxin, IL-6, fibroblast growth factor (FGF), platelet-derived growth factor (PDGF) and transforming growth factor-β (TGF). Although monocytes and macrophages are, in fact, sources of these cytokines, some are prominent products of platelets, fibroblasts, keratinocytes, and endothelial cells. In general, the various cytokine polypeptides possess diverse biological properties, most of which are associated with either host responses to various disease states or participate as part of a pathological process.

IL-1 was originally described in the 1940s as a heat-labile protein found in acute leukocytic exudate fluid which, when injected into animals or humans, produced fever. This material was a small protein (10–20 kd) and was called endogenous pyrogen [1]. In the 1970s it was shown that 30–50 ng/kg of homogeneous endogenous pyrogen produced monophasic fever in rabbits [2, 3]. No amino acid sequence for endogenous pyrogen was known, but studies demonstrated that endogenous pyrogen did more than cause fever. When injected into animals it induced hepatic acute phase protein synthesis, caused decreases in plasma iron and zinc levels, produced neutrophilia, stimulated serum amyloid A protein synthesis, and augmented T-cell responses to mitogens and antigens in vitro [4, 5]. The multiple biological activities of endogenous pyrogen, particularly its ability to affect immunocompetent cells, resulted in changing its name to IL-1. The name IL-1 now includes several substances originally named for their biological activities, but following data generated with recombinant IL-1 or compared to N-terminal amino acid sequences, these are identical to IL-1. These are leukocytic endogenous mediator [4], lymphocyte-activating factor [6], mononuclear cell factor [7], catabolin [8], osteoclast-activating factor [9], and hemopoietin-1 [10, 11].

The various cytokines listed above are for the most part structurally distinct with the exception of IL-1α/IL-1β and FGF. IL-1α and IL-1β, despite distinct primary amino acid sequences, are structurally related as shown by molecular modeling, crystallographic analysis and receptor recognition. A similar case exists for TNF and lymphotoxin. In addition, IL-1β is structurally related to fibroblast growth factors (acidic form) and shares the growth-promoting properties of these molecules [12]. TNF and IL-1 induce nearly identical biological effects, particularly those associated with systemic and local inflammatory as well as destructive joint disease [7, 13–15], but share only 3% primary amino acid structure. Receptors for IL-1 and TNF are distinct.

Considerable interest has focused on IL-1, TNF, and IL-6 as mediators of systemic 'acute phase' responses. Injecting experimental animals with either IL-1 or TNF results in fever, hypozincemia, hypoferremia, increased hepatic acute phase protein synthesis, and other manifestations of the response. IL-6 induces fever and hepatic acute phase protein synthesis. Recent evidence suggests that IL-6, like IL-1 and TNF, is present in human body fluids associated with inflammatory and febrile diseases. This chapter focuses on IL-1, but the related cytokines, TNF and IL-6, are also discussed, particularly as these molecules share biological properties with IL-1.

Interleukin-1 Structure

Two forms of IL-1 have been cloned: IL-1β was cloned from human blood monocytes [16] and IL-1α was cloned from the mouse macrophage line P388D [17]. Subsequent to the description of cDNAs to these two forms, IL-1β has been cloned in the cow, rabbit, rat and mouse and IL-1α in the human, rat and rabbit. It is unclear whether more than these two gene products exist for IL-1. IL-1β is the prominent form of IL-1 and the amount of IL-1β mRNA found in activated cells is usually 10- to 50-fold greater than the α form. In addition, culture supernates and various human body fluids contain more IL-1β than the α form. However, several studies have shown that IL-1β is readily secreted from activated cells whereas IL-1α remains cell associated.

Originally identified as a pI 7 (IL-1β) and pI 5 (IL-1α) species on isoelectric focusing, the two forms of IL-1 are initially synthesized as 31-kd precursor polypeptides and share only small stretches of amino acid homology (26% in the case of the two human IL-1 forms). Neither form contains a signal peptide sequence which would indicate a cleavage site for the N-terminus. This fact makes IL-1 a highly unique substance. Other cytokines such as TGF and TNF have clearly identifiable signal peptide sequences. Lacking a clear signal peptide, a considerable amount of the IL-1 that is synthesized remains cell associated [18]. In fact, membrane-associated IL-1 is biologically active and may be the form which participates in activating lymphocytes, particularly in lymphoid tissue where lymphocytes form rosettes around macrophages [19]. 'Membrane-bound' IL-1 is also active on nonlymphocyte target cells. The steps involved in transcription, translation, and 'processing' of IL-1 are discussed below.

Within the various animal species of IL-1β, the primary amino acid sequences are conserved in the range of 75–78% whereas the α sequences are in the range of 60–70%; between the β and α IL-1 within each species, conserved amino acid homologies are only 25%. The entire human genes for each IL-1 form have also been cloned [20, 21]. Each gene contains 7 exons coding for the processed IL-1 mRNA and raise the possibility of alternate RNA processing. The gene for human IL-1β is located on chromosome 2 [22] and the gene for mouse IL-1α is also on chromosome 2. The existence of other IL-1 forms (as separate gene products or the result of processed mRNA) has recently been introduced in studies on IL-1 from Epstein-Barr virus-infected human B cells [23a]. However, it remains to be shown whether these latter cells produce a different IL-1 gene product or that posttranslational processing results in IL-1s of different molecular weights and charges. The N-terminal amino acid sequence reported for B-cell IL-1 [23a] is unrelated to the N-termini of IL-1β or α. Recently, a T-cell factor with physical characteristics similar to those of the human B-cell IL-1 has been shown to possess the identical N-terminal amino acids [23b]. This T-cell factor is biologically related to IL-1 in that it induces the p55/TAC antigen on T cells. The factor is derived from human adult T-cell lymphotropic virus-1 transformed T-cell lines. It appears that the IL-2R-inducing lymphokine is the same as B-cell IL-1.

When the amino acid sequences of the two IL-1 forms are compared, only 4 small regions of amino acid homologies exist. These have been identified and called regions A–E [24] (fig. 1); since regions A and B are contained in the precursor sequence which is missing in the mature IL-1 form, the important regions of homology for the two mature IL-1 forms are located in the carboxyl C-D and E regions. These regions may represent a putative 'active site' of the IL-1 molecule which would explain the observation that although the two forms are structurally distinct, they share the same spectrum of multiple biological properties and recognize the same receptor. Of interest to the evolution of IL-1 is that the C-D region of IL-1β/α homology is coded by the entire VIth exon and this region also contains some limited amino acid homology with interferon-α_2 and interferon-β_2. There is also a small (3%) region of amino acid homology with TNF but this region does not correspond to the IL-1β/α homologous regions.

Since both mature forms of IL-1 recognize the same cell receptor and both forms possess the same biological properties, attention has focused on the concept that the C-D contains the minimal structural requirements for receptor activation. Several peptides have been produced either by synthetic [25] or recombinant DNA methods [26]. Although these IL-1 peptides have some biological activity, the specific activities are low and they do not block the receptor binding of the mature peptide. Antibodies to this C-D region do not block the activity of the mature IL-1β peptide on a variety of target cells,

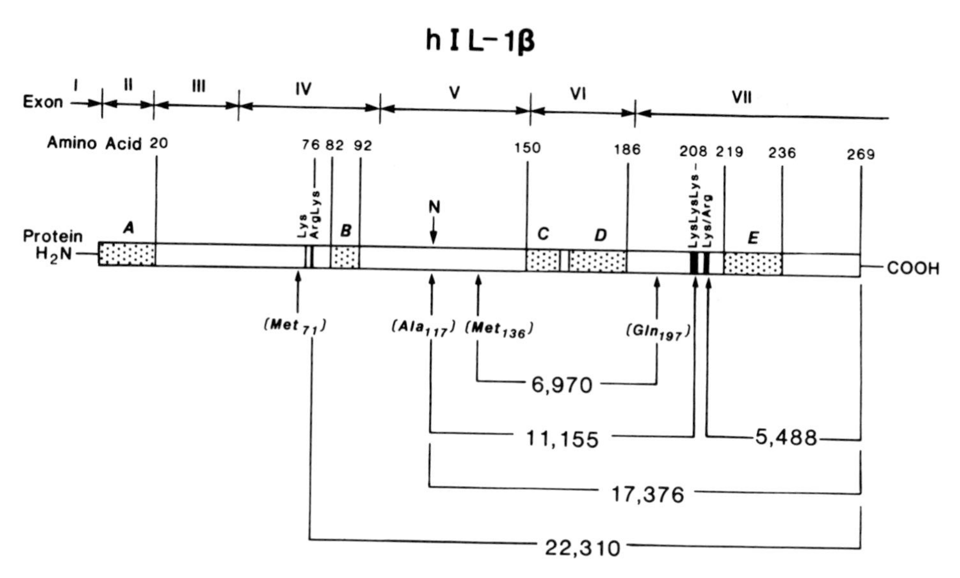

Fig. 1. Structure of IL-1β.

whereas antibodies to synthetic C-terminal and N-terminal peptides do reduce biological activity [27]. These data support other studies (see below) suggesting that both the N-terminal as well as C-terminal amino acids are involved in receptor-binding events.

Small molecular weight peptides at 4 and 2 kd, with IL-1-immunoreactive and biological activity, have been consistently isolated from human body fluids [28–30]. These small molecular weight peptides have been observed in preparations of recombinant IL-1 and appear to be generated by trypsin-sensitive sites. The amino acid sequence of human IL-1β contains several cleavage sites for serine proteases which would generate peptides of various molecular weights. These are illustrated in figure 1. There is a 6,970-dalton peptide which contains the C-D region and is generated by serine proteases. There is also a 5,488-dalton C-terminal peptide generated at the lysine-lysine-lysine site which could represent the active C-terminal fragment [31]. Inhibitors of serine proteases prevent the appearance of this and other small molecular weight immunoreactive IL-1 peptides in supernates of human monocytes [32]. Recently, an endopeptidase which is membrane bound and is involved with degradation of neuropeptides has been shown to destroy the biological activity of IL-1β [33].

Human IL-1β has recently been crystallized and its tertiary structure analyzed at a resolution of 3.0 Å [34]. The molecule that was crystallized had an N-terminus at Ala_{117} and hence represents a 17.5 kd processed form. This is the dominant form found in extracellular fluids. The three-dimensional structure revealed 12 β strands forming a complex of hydrogen bonds. Computerized molecular modeling of IL-1 from the primary sequence also revealed a similar structure [35]. The basic structure is similar to a tetrahedron, the interior of which is filled with hydrophobic side chains. The overall folding of the 12 β strands is similar to that found in the soybean trypsin inhibitor. The interior of the IL-1β is strongly hydrophobic with no charged amino acids in the interior. The histidine at position 147 is on the surface of the molecule and when this amino acid is substituted by site-specific mutation, a corresponding loss of biological and receptor-binding activity takes place [36]. The N-terminal mutations have also yielded altered biological and receptor-binding data [37], suggesting that N-terminal amino acids, seemingly the result of limited proteolysis by serine proteases [38], play an important role in either stabilizing the tertiary structure or by direct interaction with receptor-binding domains. Studies on the active site of IL-1 have been carried out using antibodies directed against different synthetic peptides and assessing their ability to neutralize the immunostimulatory as well as inflammatory properties of the whole IL-1 [27, 39].

FGFs have been shown to stimulate fibroblast and endothelial cell proliferation, smooth muscle cell proliferation and angiogenesis, properties similar to those of IL-1. Like IL-1, these molecules exist in two forms, acidic (pI 5) and basic (pI 8). Bovine brain-derived acidic and basic FGFs have significant amino acid homologies with the IL-1β [12] and to a lesser degree with IL-1α. The stretches of IL-1 and FGF amino acid homologies are distributed throughout the sequences. However, analysis of the IL-1 sequence in the C-D region (see above) with that of the FGFs does not reveal any particular homology in this region. Nevertheless, the amino acid sequence similarities between the two IL-1 forms and the two forms of FGF support the observation that some biological properties are shared, particularly their ability to induce cell proliferation. It is unknown whether IL-1 binding to fibroblasts is displaced by FGFs. Other growth factors such as TGF (β and α) and PDGF do not have any structural homologies to IL-1 or TNF at the level of amino acid sequences. In addition, the receptors to these cytokines appear distinct. It seems likely that these different cytokines bring about similar biological changes because they induce similar postreceptor cellular signals.

Gene Expression, Synthesis and Processing of IL-1

A critical aspect of understanding IL-1 gene expression in a variety of cells is the exquisite sensitivity of some IL-1-producing cells to the effects of endotoxins (bacterial lipopolysaccharides). This is particularly the case with human blood monocytes which will produce IL-1 when stimulated by concentrations of endotoxin as low as 5–10 pg/ml. Routine tissue culture media contain orders of magnitude greater amounts of endotoxin. It is often difficult to assess gene expression in some experiments since IL-1 transcription can easily be stimulated by routine laboratory culture media. In order to demonstrate increases in IL-1 mRNA, some investigators have used 10 μg/ml of endotoxin to show stimulation of IL-1 transcription over that of the 'unstimulated' control [40]. Therefore, in many studies, IL-1β transcription has already taken place during the preparation and early culture of monocyte/macrophages. Adherence to glass and some plastic surfaces can serve as a stimulus of RNA synthesis. If one carefully separates human blood mononuclear cells on endotoxin-free Ficoll-Hypaque, avoids activating cells by adherence or endotoxin-contaminated culture media, there is no IL-1 mRNA present in unstimulated cells and no IL-1 protein is translated, including both intracellular and extracellular compartments [41].

Transcription of IL-1 mRNA is rapid in stimulated cells: in both human macrophage cell lines as well as in human blood mononuclear cells, endotoxin-stimulated IL-1β RNA transcription can be observed within 15 min [42, 43]. In human cultured endothelial and smooth muscle cells, a similar rapid increase has been reported [44, 45]. Transcription increases and reaches peak levels in 3–4 h and then levels off for several hours before decreasing. Transcription of IL-1α appears to be under tighter control in that inhibitors of protein synthesis are sometimes required to observe IL-1α RNA [45]. In fact, the total amount of IL-1β mRNA increases and is maintained at higher levels when inhibitors of protein synthesis are used [42]. During endotoxin stimulation, transcriptional repressors are translated (or activated by phosphorylation) and these either suppress further transcription or increase mRNA degradation.

In addition to a tight control over transcription, IL-1 is translated by a mechanism which is poorly understood. For example, human monocytes can be stimulated by adherence to glass surfaces and yet not translate any of the mRNA into IL-1 protein. In fact, the level of mRNA following adherence to glass or cellulosic membranes can be as high as that following endotoxin stimulation and yet no IL-1 protein is produced in the absence of another stimulant, in most cases endotoxin. This requirement for a second signal for translation is similar to the case for ferritin biosynthesis in which cells contain high levels of mRNA but require iron to stimulate translation. Of course, endotoxin and similar stimulants serve the dual purpose of initiating transcription as well as translation. IL-1 itself serves to stimulate both transcription and translation of IL-1 [46, 47]. Corticosteroids, when added to cells before they are stimulated, prevent both transcription as well as translation [48]. However, if corticosteroids are added after IL-1 mRNA is present, there is no evidence of decreased transcription and most of the effect is on translation [40].

Prostaglandins and prostacyclins have no effect on transcription but reduce translation [48, 49]. Blocking cyclooxygenase results in increased production of IL-1 protein, particularly when cells are stimulated by agents which increase PGE synthesis. In these situations, reduction of PGE and PGI by cyclooxygenase inhibitors removes the suppressive effect of the arachidonate metabolites, but this effect may represent the artifact of in vitro cultured cells since PGE accumulates under these conditions whereas in vivo, efficient mechanisms exist to rapidly remove PGE metabolites. The mechanism of PGE-induced suppression of IL-1 translation appears to be via the induction of cAMP [50]. The addition of PGE and dibutyl cAMP or PGE and theophylline augments the suppression. It is unclear exactly how increased cAMP affects the translation of IL-1 protein but studies show no effect on IL-1β mRNA when cAMP degradation is prevented by inhibitors of phosphodiesterase.

Despite the lack of a signal peptide, IL-1 is found in the supernates of stimulated monocytes and macrophages. Other cells producing IL-1, for example, endothelial cells, keratinocytes, smooth muscle cells, and renal mesangial cells, transcribe large amounts of IL-1 mRNA and translate IL-1 protein, but a considerable amount of the IL-1 remains intracellular as the precursor molecule (31 kd) [32, 51]. The amount of IL-1 that is 'secreted' depends upon the cell type and the conditions of stimulation. The monocyte/macrophage appears to be the cell best equipped to 'secrete' IL-1. These cells contain polyadenylated RNA coding for IL-1β at concentrations as high as 2–5% of the total poly A after adherence and stimulation by endotoxin. Most of the IL-1β mRNA transcribed is translated under these conditions. Using radioimmunoassays or enzyme-linked assays, studies indicate that as much as 100 fg of IL-1 is synthesized per human monocyte (or 100 ng/10^6 monocytes) during the 24 h following stimulation [52, 53]. Although the amount of mRNA coding for IL-1α is approximately 20–50% less than IL-1β in these cells [54], there is more total (cell-associated plus extracellular) IL-1α protein produced following endotoxin stimulation [53, 55, 56].

The reason for this discrepancy between the amount of IL-1α mRNA and the amount of IL-1α protein translated remains unclear. One possibility is that a considerable amount of the IL-1β mRNA is never translated whereas translation of the IL-1α RNA is highly efficient. An alternative explanation is that the mRNA for IL-1β is more rapidly degraded than that for IL-1α. Control of IL-1 translation is affected by other cytokines. IL-1-induced IL-1 production (either IL-1β-induced IL-1α or IL-1α-induced IL-1β) is suppressed by γ-interferon, whereas γ-interferon augments the amount of IL-1 synthesized following endotoxin or TNF stimulation [57]. The suppression of IL-1-induced IL-1 by γ-interferon is posttranscriptional. Posttranscriptional suppression of IL-1 synthesis is also observed in cells treated with PGE_2 and is a cAMP-dependent mechanism [50]; corticosteroids can suppress IL-1 synthesis when added before transcription [48] as well as after transcription [40].

As depicted in figure 2, activators of cells for IL-1 synthesis also trigger the events leading to increased prostaglandins and leukotrienes. As mentioned above, prostaglandins suppress IL-1 translation; however, leukotrienes appear to augment IL-1 production. This has been shown by adding LTB-4 to human monocytes and stimulating IL-1 production [58]. Agents that block the lipoxygenase pathway of arachidonate metabolism leading to formation of leukotrienes also reduce IL-1 production [49, 59]. Similar studies demonstrate that this series of events also occurs in macrophages producing TNF. Recent evidence supports the importance of lipoxygenase products in the production of IL-1. In human volunteers taking eicosapentaenoic acid fatty acid dietary supplements, there is a 60% reduction in the ability of their mononuclear cells to synthesize IL-1β and IL-1α in vitro [56]. A similar observation was made for TNF. The mechanism probably involves the ability of these omega-3 fatty acid precursors to be metabolized to LTB-5 rather than LTB-4. LTB-5 competes with LTB-4 for receptor occupancy. It is unclear at which stage the lipoxygenase metabolites act on IL-1 production. Since one can add a lipoxygenase inhibitor 1–2 h after cell stimulation without affecting the amount of IL-1 synthesized, it appears that lipoxygenase metabolites are involved with early events such as transcription.

The first translation product of IL-1 is the 31-kd precursor. This can be found mostly in the intracellular pool. The intracellular pool also contains other molecular weight fragments of IL-1 at 22 and 17 kd, but it is unclear whether these occur as a result of artifactual proteolysis during specimen

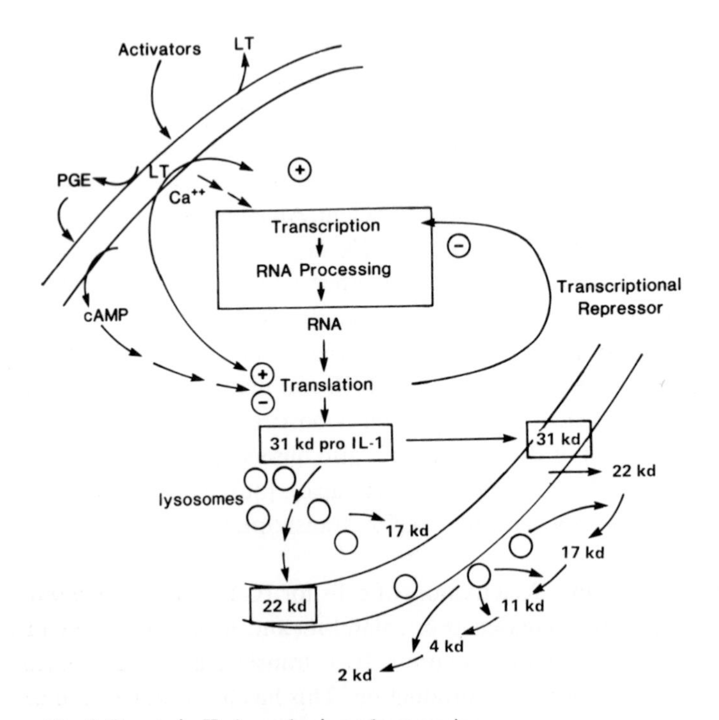

Fig. 2. Events in IL-1 synthesis and processing.

preparation [32]. The localization of cell-associated IL-1 (as compared to extracellular IL-1) is almost entirely cytoplasmic. Using antibody staining or radioimmunoassays, the data consistently indicate that cell-associated IL-1β is primarily in the cytosol, not in the endoplasmic reticulum, Golgi, or plasma membrane fraction [60, 61]. Despite the failure to measure IL-1 in the plasma membrane fractions, several studies have shown IL-1β staining on the cell surface [62]. Lysosomal localization appears to be important for the processing and secretion of IL-1 [60, 63]. Studies on the localization of IL-1α also indicate a similar predominance in the cytosol [61], but others have demonstrated IL-1α associated with the plasma membrane [64]. IL-1α is also phosphorylated and it is unclear how this contributes to its cellular localization [65].

There is a 22-kd form which can be isolated from the intracellular pool and the extracellular fluid. The 22-kd IL-1 is also thought to be an intermediate form which may be located transiently in the membrane. The 31-kd IL-1 precursor is immunoprecipitated by anti-IL-1α where it is biologically active as an inducer of hepatic acute phase protein synthesis [64]. Most of the IL-1 found in the extracellular fluid, however, is the 'mature' 17.5-kd peptide with an N-terminal at position 117 (alanine) for the IL-1β precursor [66] and at position 113 (serine) for the human IL-1α. Smaller peptide fragments of these mature peptides have been found in monocyte supernates and the appearance of these peptides can be reduced in the presence of serine protease inhibitors [32]. Elastase and plasmin have been implicated as monocyte proteases which cleave IL-1 into its 17.5-kd mature fragment [51]. This and other various subfragments are biologically active and are routinely found in human plasma, urine, peritoneal, pleural, and joint fluids. As shown in figure 2, smaller peptides with molecular weights at 22, 17, 11, 6, 4 and 2 kd are found in the extracellular fluid of stimulated human monocytes and may correspond to the potential cleavage sites depicted in figure 1.

The mechanism of IL-1 cleavage is thought to be via lysosomal enzymes [63]. There is a correlation between the amount of processed 17.5-kd mature IL-1 found in the extracellular fluid and the type of stimulus used to activate the monocyte or macrophage. Particles such as *Staphylococcus epidermis* or zymosan induce large amounts of IL-1 that is mostly mature and extracellular. These stimulators are also potent inducers of lysosomal exocytosis. Other stimulators such as a low dose (50–100 pg/ml) of bacterial endotoxins induce approximately equal amounts of intracellular and extracellular IL-1β whereas IL-1α is nearly entirely cytosolic. Adherence to plastic or glass surfaces in the absence of endotoxins induce mRNA for IL-1β but no detectable IL-1 protein. Small amounts of endotoxin rapidly induce translation of this mRNA [41]. However, most commercial tissue culture media or sera contain sufficient endotoxin concentrations (10–50 pg/ml) to stimulate cell-associated IL-1, including membrane-bound IL-1 which is biologically active [19]. Thus, it is likely that most cultures containing monocyte/macrophages have active cell-associated IL-1 despite a failure to demonstrate extracellular IL-1.

The Biological Effects of IL-1

The expression of recombinant IL-1 has been accomplished and, in general, there does not seem to be any difference in the spectrum of biological activities of either form. IL-1β is vulnerable to oxidation, and thus biological specific activities of IL-1β can be lower than those of IL-1α which is highly stable. If IL-1β is purified under nondenaturing conditions, it is equipotent with the α form in a variety of in vivo and in vitro asssays. The binding of either IL-1β or IL-1α to various cells in receptor binding assays is blocked by each form [67]. Both recombinant human IL-1β and α augment T, B and natural killer (NK) cell responses. The ability of IL-1 to activate immunocompetent cells seems unique to the group of cytokines which affect cellular growth and proliferation; FGF, PDGF and TGF have either no effect on immunocompetent cells, or in the case of TGF-β, are potent immunosuppressive agents [68]. Both forms of IL-1 induce systemic acute phase responses including fever, hepatic acute phase protein synthesis, neutrophilia, hypoferremia, hypozincemia, increased levels of hormones, and cause sleep. At higher doses, IL-1 induces hypotension and a shock-like state [69]. Some biological properties reported for natural IL-1 have not been confirmed with either recombinant form. These include the ability of IL-1 to cause neutrophil superoxide production and degranulation in vitro [70] and to induce muscle proteolysis in vitro [71, 72]. Others, however, do show an effect of recombinant IL-1β on muscle proteolysis [73], particularly a fragment of IL-1 generated from the recombinant IL-1β form. It appears IL-1 effects on neutrophils is due to its ability to augment the biological effects of other neutrophil activators such as the chemotactic peptides. There are receptors for IL-1 on neutrophils [74], and IL-1 acts as a cofactor or permissive factor for the activity of f-met-leu-phe. Recombinant human IL-1, however, directly stimulates basophil [75, 76] and eosinophil degranulation of histamine and aryl sulfatase [77], respectively.

The multiple biological activities of IL-1 have been studied in terms of in vivo and in vitro effects. In patients with bacterial infection, injury or chronic inflammatory disease, IL-1 may account for a majority of observed acute phase changes. It is difficult to study such subjective symptoms as headache, myalgias, arthralgias and lassitude in animal models, but the potency (10^{-12}–10^{-15} M) of IL-1 in inducing release of prostaglandin E_2 (PGE_2) from fibroblasts, synovia and other cells suggests that these symptoms are likely mediated by increased levels of IL-1.

Effects on Hepatic Protein Synthesis

Hepatic proteins which change during acute phase responses include clotting factors, complement components, fibrinogen, haptoglobin, ceruloplasmin and others. In addition, there are increases in hepatic proteins generally not synthesized in health but in association with infection, injury or other pathological processes. IL-1, TNF and hepatocyte-stimulating factor (now identified as the same molecule as interferon-β_2 and B-cell-stimulating factor-2 or IL-6) play important roles in regulating the synthesis of hepatic proteins. The cytokine-induced increases in normal hepatic proteins are usually in the range of 2- to 3-fold, but the synthesis of pathological proteins can increase 100- to 1,000-fold. Two such proteins, serum amyloid A (SAA) protein and C-reactive protein (CRP) are classical 'acute phase reactants', and serve as markers of disease. SAA contributes to the development of secondary amyloidosis.

IL-1 induces hepatocytes to synthesize a spectrum of acute phase proteins; these include SAA [78], fibrinogen [79], complement components and various clotting factors. At the same time, albumin levels decrease. Studies have shown that IL-1 regulates the synthesis of these and other hepatic acute phase proteins at the level of mRNA transcription. In isolated hepatocytes, IL-1 decreases the transcription of RNA coding for albumin, increases transcription of factor B, initiates SAA mRNA synthesis, but has no effect on gene expression of a control protein, actin. Other proteins have been studied in hepatic cell line cultures (HepG2 and Hep3B), and in picomolar ranges, IL-1 stimulates the biosynthesis of complement protein C3 and α_1-antichymotrypsin [80]. There is also a modest stimulation of α_1-acid glycoprotein and inter-α_1-trypsin inhibitor synthesis. In addition to decreased albumin transcription, hepatic cells exposed to IL-1 synthesize less transferrin. In hepatic cells lines, IL-1 does not increase the expression of CRP, although the intravenous injection of recombinant IL-1 does result in elevated CRP levels after 24 h. In murine fibroblasts transfected with cosmid DNA bearing the genes for C2 and factor B, IL-1 stimulated the expression of factor B but did not affect synthesis of C2 [81].

IL-1 has other effects on liver metabolism. It depresses the activity of liver cytochrome P-450-dependent drug metabolism in mice [79] and this observation may explain the impaired drug clearance and excretion in patients with infections and fever. The liver's response to IL-1 also includes the synthesis of metalloproteins which bind serum iron and zinc and account for the hypozincemia and hypoferremia induced by IL-1. Bacteria and tumor cells require large amounts of iron for cell growth, particularly at elevated temperatures, and the ability of the host to remove iron from tissue fluids seems to be a fundamental host-defense mechanism.

Effects of IL-1 on Endothelial Cells

Of its many biological properties, IL-1-induced changes in endothelial cells relate directly to the initiation and progression of pathological lesions in vascular tissue. From a physiological viewpoint, IL-1 activates human endothelial cells in vitro to synthesize and release PGI_2, PGE_2, and platelet-activating factor [82, 83]. A 10-fold increase in PGI_2 release is observed with concentrations of IL-1 in the femtomolar range. Although arachidonate metabolites increase blood flow, IL-1 also orchestrates a cascade of cellular and biochemical events that lead to vascular congestion, clot formation, and cellular infiltration. One of these initiating steps involves the ability of IL-1 to alter endothelial cell plasma membranes so that neutrophils, monocytes, and lymphocytes adhere avidly [84]. Endothelial cells need to be exposed to IL-1 for 1 h or less in order to increase their adhesiveness. The action of IL-1 in this process appears to be related to the interaction of the leukocyte-glycoprotein complex called 'leukocyte function antigen' with a fibroblast and endothelial cell surface molecule called 'intercellular adhesion molecule-1'. Within 1 h following IL-1 exposure, endothelial cells increase their expression of intercellular adhesion molecule-1 [85]. Patients with defective leukocyte function antigen expression have repeated bouts of bacterial infection. In addition to activating endothelial cell-leukocyte adhesion, IL-1 also increases the binding and lysis by NK cells by a variety of tumor targets and is chemotactic for monocytes and lymphocytes. Consistent with the effects of IL-1 on leukocyte chemotaxis and adherence to endothelial cells, IL-1 injected intradermally causes the accumulation of neutrophils and IL-1 can substitute for endotoxin in either limb of the local Shwartzman reaction. This latter property is most apparent using the combination of IL-1 and TNF [86].

IL-1 increases endothelial cell surface procoagulant activity [87], and production of a plasminogen activator inhibitor [88]. These events lead to activation of thrombin in the initiation of clotting. Taken together, these effects would decrease blood flow in vessels and increase the accumulation of leukocytes and platelets. Since IL-1 stimulates neutrophil thromboxane release [89], activated neutrophils adhering to endothelial cells may increase platelet aggregation. Finally, IL-1 has angiogenic properties in the rabbit eye anterior chamber model [Prendergast and Dinarello, unpubl. observations] and following brain injury [90]; this may be related to the fact that IL-1 and the related fibroblast growth factors share significant amino acid homologies. In general, the effects of IL-1 on endothelial cells represent a well-coordinated effort to localize tissue inflammation and contribute to the initiation of pathological lesions leading to vasculitic-like changes. Figure 3 illustrates the activity of IL-1 (and TNF, see below) on vascular tissue.

The effects of IL-1 stimulation of endothelial cell functions should be considered in light of the fact that endothelial cells produce their own IL-1.

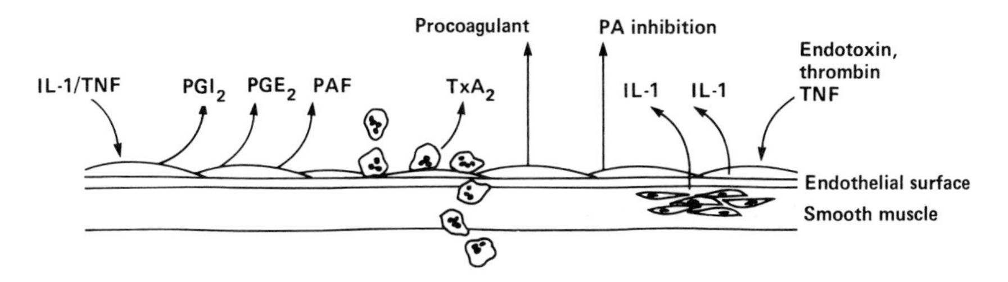

Fig. 3. IL-1 and TNF effects on vascular tissue.

Nanogram/milligram concentrations of bacterial endotoxins or TNF induce cultured endothelial cells to release IL-1 [44]. In addition, thrombin stimulates endothelial cell IL-1 production. Northern hybridization of endothelial cell mRNA supports the close relationship between the predominant IL-1β and endothelial cell-derived IL-1. Thus, the induction of endothelial cell IL-1 by two clinically relevant stimulators (endotoxin and thrombin) may initiate a cascade of events leading to further development of vasculitic processes. Although immune complexes stimulate monocyte IL-1 production, there are no reported studies as yet demonstrating that immune complexes stimulate endothelial cell IL-1 synthesis. Recent studies demonstrated that arterial smooth muscle produces IL-1 [45], that IL-1 is a growth factor for smooth muscle cells, and IL-1 induces IL-1 in these cells [47].

Catabolic Effects of IL-1

The catabolic properties of IL-1 are usually considered in terms of its local effects. For example, IL-1 produced locally acts in a paracrine-like fashion in destructive joint and bone disease and local tumor invasion. On the other hand, IL-1 in the systemic circulation exerts its catabolic effects indirectly by affecting the metabolism of lipoproteins by liver and fat. IL-1 is a potent inducer of collagenase production in synovial cells [7, 91]; in addition, IL-1 induces release of metalloproteinases and proteoglycanases from chondrocytes [8, 92]. In fact, because of its local catabolic biological activites, pig IL-1 was previously known as 'catabolin'. Recombinant IL-1s added to bone cultures in vitro induce dramatic resorptive processes and shrinkage of bone matrix. Just as pig IL-1 was known previously as 'catabolin', osteoclast-activating factor (OAF) is now identified as having the same amino acid sequence as the human IL-1β [9]. Thus, it is presently considered that the catabolic properties of IL-1 in cartilage and bone contribute to the tissue destruction and matrix loss in a variety of joint diseases [7, 93]. As discussed below, TNF can also act as an OAF. IL-1 and the related FGF synergize in the induction of proteases from chondrocytes [94].

Effects of IL-1 on Fibroblasts and Fibrosis

In contrast to its catabolic activities, IL-1 increases fibroblast proliferation [95] and collagen synthesis [96]. IL-1 is mitogenic for fibroblasts and is thought to play a major role in pannus formation. In fact, cellular infiltration and early pannus formation has been observed in rabbit joints injected with recombinant IL-1 [93]. In vitro, however, IL-1-induced fibroblast and smooth muscle cell proliferation is often difficult to observe unless cyclooxygenase inhibitors are present [44, 45]. This seems to be related to the PGE_2 levels induced by IL-1. This is not the case with other FGFs such as PDGF, epidermal growth factor and TGF. In comparison to epidermal growth factor and TGF-β, numbers of receptors for IL-1 on fibroblasts are low (2,000/cell) [97, 98], but compared to IL-1 receptors on other cells, the fibroblast ranks as one of the tissues with high numbers of receptors. One mechanism for IL-1-induced fibroblast growth is IL-1 via increased production of receptors for endogenous growth factors such as epidermal growth factor. In terms of IL-1's role in fibrosis, recombinant IL-1 directly increases the transcription of type I and type III collagen [96] and type IV (basement membrane) collagen [99].

Other Cytokines Affecting Fibroblast and Fibrosis

Other cytokines share with IL-1 the ability to stimulate fibroblast proliferation. These include TNF [100], FGF [101], PDGF [102], and TGF-β. In addition, another growth factor from macrophages, called macrophage-derived growth factor, seems to act as a competence factor for PDGF-induced fibroblast growth [103]; however, this factor now seems to be PDGF itself. The mechanism of action of PDGF on cell proliferation includes a requirement for insulin whereas IL-1 and TNF effects are seen in the absence of insulin. Despite the fact that these various cytokines have distinct receptors, their effects on cell growth appear similar. These cytokine growth factors also share other biological properties. For example, like IL-1, PDGF is chemotactic for neutrophils and monocytes in vivo, induces cell aggregation and adhesiveness and induces degranulation [104, 105]. IL-1, TNF, TGF, and PDGF induce the synthesis of collagen and collagenases. At present, the multiple biological properties of IL-1 are clearly shared, to some

Table 1. Systemic effects of recombinant IL-1[1]

Central nervous system	*Metabolic*
Fever	Hypozincemia, hypoferremia
Brain PGE_2 synthesis	Decreased cytochrome P-450 enzyme
Increased ACTH	Increased acute phase proteins
Decreased REM sleep	Decreased albumin synthesis
Increased slow wave sleep	Increased insulin production
Decreased appetite	Inhibition of lipoprotein lipase
	Increased sodium excretion
	Increased corticosteroid synthesis
Hematologic	*Vascular wall*
Neutrophilia	Hypotension
Nonspecific resistance	Increased leukocyte adherence
Increased GM-CSF	Increased PGE synthesis
Radioprotection	Decreased systemic vascular resistance
Bone marrow stimulation	Decreased central venous pressure
Tumor necrosis	
	Increased cardiac output
	Increased heart rate
	Decreased blood pH
	Lactic acidosis
	Chemoattractant

[1] These effects have been demonstrated by administration of recombinant IL-1 to animals.

extent, with these growth factors, raising issues such as synergism and antagonism between the cytokines on mesenchymal tissue. Recent data demonstrate that considerable synergism between IL-1 and TNF exist in a variety of biological effects; little is known about the synergism or antagonism between the non-IL-1 growth factors, except that TGF-β is immunosuppressive [68].

Systemic versus Local Effects of IL-1

Table 1 lists the systemic effects of recombinant IL-1. These studies were carried out by injecting IL-1 into experimental animals either intravenously or intraperitoneally. In many cases, the IL-1-induced changes mimic the animal's responses to an injection of microbial toxins or antigen-antibody complexes. When administered intravenously into rabbits at 200 ng/kg, human recombinant IL-1 results in a fall in circulating neutrophils within 5 min and this is due to rapid margination on the endothelial surfaces; approximately 10 min after an intravenous injection, rectal temperature begins to rise and reaches peak elevation between 45 and 55 min after the injection [38]. Increases in slow wave sleep parallel the fever course but the sleep-inducing property of IL-1 is not linked to its ability to increase body temperature. After 4–6 h, temperature and sleep patterns return to preinjection levels but there is an increase in circulating neutrophils, particularly new forms released from the marrow. After 12 h, serum albumin, iron, zinc and cytochrome P-450 enzyme activity are decreased [79]. When approximately 10 times the amount of IL-1 is injected, a decrease in systemic arterial pressure, systemic vascular resistance, and central venous pressure can be observed within 10 min [69]. Although these hemodynamic changes are similar to the changes observed in animals given TNF, tissue damage is not characteristic of IL-1, at least at doses less than 5 μg/kg, whereas TNF is associated with death in some organs [106]. Large doses of recombinant IL-1 injected into mice have been given without death or tissue damage unless the mice have been previously adrenalectomized [P. Ghezzi, pers. commun.].

Information derived from the in vitro effects of IL-1 may relate to some of the biological properties of IL-1 observed in vivo. For example, IL-1 stimulates somatostatin release from cultured pituitary cells [107] and also increased steroid synthesis [108] from perfused adrenal glands. Therefore it is not surprising that other neuropeptides are stimulated when IL-1 is given systemically. The effects of IL-1 on endothelial cell arachidonic acid metabolism, platelet-activating factor generation, and procoagulant activity in vitro likely explain the shock-like state that IL-1 produces in vivo [69]. In addition, the ability of IL-1 to induce various lymphokines in vitro, for example, interferon-β_1, interferon-γ, IL-1, IL-3 and IL-6, likely occurs in vivo, but it is difficult to demonstrate circulating levels of interferons and interleukins unless large doses of IL-1 are administered. One approach to interpreting the large body of evidence for the multiple biological effects of IL-1 is to view local production and action as the 'autocrine/paracrine' action of IL-1. These are listed in table 2. By contrast, the systemic effects of IL-1 can be viewed as the 'hormonal' property such as fever and decreased appetite [109]. At present, it seems that autocrine/paracrine effects of IL-1 predominate in some diseases such as type I diabetes mellitus [110], whereas systemic effects are characteristic of IL-1 produced as a result of toxemia, septicemia, widespread tissue damage or intravenous antigen challenge.

Table 2. Local effects of recombinant IL-1[1]

Nonimmunological
Chemoattractant (in vivo)
Basophil histamine release
Eosinophil degranulation
Increased collagenase production (in vivo)
Chrondrocyte protease release
Bone resorption
Induction of fibroblast and endothelial CSF activity
Production of PGE_2 in dermal and synovial fibroblasts
Increased neutrophil and monocyte thromboxane synthesis
Cytotoxic for human melanoma cells
Cytotoxic for human islet cells
Cytotoxic for thyrocytes
Keratinocyte proliferation
Proliferation of dermal fibroblasts
Increased collagen synthesis
Mesangial cell proliferation
Gliosis
Immunological
T-cell activation
IL-2 production
Increased IL-2 receptors
B-cell activation
Synergism with IL-4
Induction and synergism with IL-6
Activation of NK cells
Synergism with IL-2 and interferon NK cells
Increased lymphokine production (IL-3, IL-6, IFN-γ)
Macrophage cytotoxicity
Growth factor for B cells
Increased IL-1 production

[1] These effects are derived from in vitro studies.

Local Effects

It is increasingly clear that tissues producing IL-1 are either themselves targets of IL-1 or are capable of acting on adjacent tissues. For example, IL-1 produced by macrophages in a lymph node acts on macrophages as well as on lymphocytes inducing IL-2, IL-2 receptors, IL-3, IL-6 and interferon-γ. IL-1 produced by microglia and astrocytes in the brain has its effect on local gliosis [111] but may induce no systemic responses. A similar case can be made for the ability of IL-1 to stimulate production of granulocyte-macrophage colony-stimulating factor (GM-CSF) [112, 113], interferon-β_1 [66] and IL-6 [114, 115] in bone marrow. The local production and biological activity of IL-1 in the joint space has attracted considerable attention because of a potential role for IL-1 in the pathogenesis of various joint diseases. However, in the joint space, IL-1 likely exerts its autocrine/paracrine effects in the absence of major systemic responses. A similar case can be made for IL-1 which is found in the nerve fibers of the human hypothalamus [116]. Locally produced IL-1 in the brain may exert various neuroendocrinologic changes with no detectable systemic changes other than what is detected as normal physiologic changes. The ability of membrane or cell-associated IL-1 to be biologically active underscores the importance of local effects of IL-1 in the absence of systemic signs and symptoms.

Effect of IL-1 on Immunocompetent Cells

Is IL-1 Required for T-Cell Activation?

Recombinant IL-1 has been used to confirm a role for IL-1 in the mechanism of T- and B-cell activation, eliminating previous doubts about the purity of natural IL-1 preparations and whether these were capable of participating in immune responses. This is a particularly important issue because recent data on the ability of IL-6, IL-4 [117] and GM-CSF to act as 'lymphocyte-activating factors' have been used to cast doubts on the role of IL-1 as an activator of lymphocytes. Clearly, some preparations of natural IL-1 could have been contaminated with other lymphokines and clouded the issue. However, recombinant IL-1 and antibodies to recombinant IL-1 or antibodies to synthetic IL-1 peptides have clarified the presence of IL-1 receptors on lymphocytes and that IL-1 acts as an activator of lymphocytes. Recombinant IL-1 has not, however, helped settle the issue of whether IL-1 is an absolute requirement for a primary immune response. Certain aspects of the biology of IL-1 have confounded the issue of whether soluble IL-1, like IL-2 and other T-cell growth factors, is required for T-cell activation. For example: (1) some T cells synthesize IL-1 [118]; (2) IL-1 induces other cytokines (such as IL-6 and γ-interferon, and (3) IL-1 is biologically active as a cell-associated protein, presumably as a cell membrane surface protein [19].

There is little doubt that in many models of T- and B-cell activation, IL-1 amplifies the response to antigens or mitogens. This is particularly the case when suboptimal concentrations (concentrations that actually may be close to the in vivo situation) are used and when some attempt at macrophage depletion has been employed. In fact, the original immunologic effect of IL-1 was its ability to act in a costimulator assay, the so-called 'lymphocyte-activating factor' assay using murine thymocytes and suboptimal amounts of mitogens. It soon became clear that given the proper conditions, IL-1 would amplify the activation and induce IL-2 gene expression and IL-2 receptors [119]. What remains to be resolved since those initial observations is whether mature, resting T cells have an absolute requirement for IL-1 during primary activation. The present data suggest that in the strict absence of macrophage membranes, soluble, recombinant IL-1 does not restore the macrophage requirement for activation leading to proliferation. In examining the role for IL-1 in T-cell activation, it appears best to divide the discussion into three areas : (1) the effect of IL-1 on thymocytes; (2) the effect of IL-1 on T-cell cell lines, and (3) the ability of IL-1 to activate mature, resting T cells.

Effect of IL-1 on Thymocyte Responses to Mitogens

From the initial experiments of Gery and Waksman [6], IL-1 had been 'defined' as a comitogen for murine thymocytes. Further work demonstrated that when depleted of macrophages, rat thymocytes (or lymph node cells) will not proliferate when incubated with concanavalin A (ConA) and are unresponsive to IL-2 but require IL-1. Alone, IL-1 does not lead to proliferation of thymocytes, an observation repeatedly made by many investigators. Although IL-1 treatment does not affect the ability of thymocytes to bind ConA, the addition of soluble, recombinant IL-1 render the thymocytes and lymph node cells capable of responding to ConA by producing both IL-2 and IL-2 receptors [120]. In a similar fashion, peanut agglutinin-negative murine thymocytes proliferated when IL-1 is added to suboptimal concentrations of IL-2. The combination of IL-1 and IL-2 induced the IL-2 receptors [121]. The monoclonal antibody directed against the thymocyte receptor L3T4 blocks the ability of IL-1 to act as a comitogen and it has been speculated that the L3T4 molecule is functionally involved with IL-1-induced thymocyte proliferation [122]. Other data show that the Ly-1 antigen may serve as an IL-1 receptor [123].

Effect of IL-1 on IL-2 Production from T-Cell Lines

The D10.G4.1 murine T-helper cell line [124] responds to IL-1 in the femtomolar range and may be an ideal cell to study the T-cell response to IL-1. Initially this cell line entered a proliferative phase when IL-1 was added to mitogen-primed cells; in the absence of IL-1, there was no proliferative signal. Subsequently, it was shown that the effect of IL-1 on these cells was not through increased production of IL-2 but rather that IL-4 was providing the proliferative signal. Several laboratories have now demonstrated the exquisite sensitivity of these cells to IL-1 (subfemtomolar concentrations) and that these cells will proliferate to IL-1 in the absence of mitogens and antigens [125, 126]. The IL-1 receptors for these cells have been demonstrated to be comprised of both a high as well as low affinity class and the numbers of high affinity receptors is in excess of 20,000 per cell [127]. At present it remains unclear whether these cells are inducing a secondary cytokine which provides the proliferative signal, such as IL-4 [128] or colony-stimulating factors [129].

There is a growing body of evidence that the largest number of high affinity receptors for IL-1 exists on T-cell lines. These include the human Jurkat cell, EL4 mouse thymoma cell line, the LBRM33 mouse thymoma cell line, the NOB-1 line (a subclone of the EL4 cell), and the D10S subclone of the murine T-helper cell line, D10.G4.1. With the exception of the NOB-1 and the D10S cells, cell lines produce little or no IL-2 in response to IL-1. However, when costimulated with mitogens, antibodies to the CD3 complex, ionophores or phorbol esters, the cells produce large amounts of IL-2. For example, the human Jurkat cell line produces no IL-2 in response to IL-1, but in the presence of mitogens and phorbol esters there is increased transcription of IL-2 mRNA [130]. Similar observations have been established in the EL4 and LBRM33 cell lines, i.e., IL-1 itself does not induce IL-2 secretion but requires a second signal, usually one which increases cytosolic calcium [131–134].

These data are consistent with the well-established synergistic effect of IL-1 with antigens, mitogens, ionophores or activators of protein kinase C and support the existence of a two-signal hypothesis for the induction of IL-2 in T cells and T-cell lines [134]. The first signal is provided by agents which increase cytosolic calcium whereas increases in protein kinase C is a signal 2 event. Thus IL-2 synthesis takes place when cells are stimulated with ionophore and phorbol esters comparable to mitogen and IL-1, respectively. Examining the events in the LBRM33 cell line, mitogen induces rapid hydrolysis of phosphatidylinositol-4,5-bisphosphate (PI), and increased cytosolic calcium. Phorbol esters induced a protracted association of cellular protein kinase C with the plasma membrane. IL-1 however, although providing the signal which lea ds to increased IL-2 production when cells are stimulated with mitogen, did not activate protein kinase C nor mobilize calcium. Clearly IL-1 in combination with mitogen-induced intracellular events was able to trigger IL-2 production by a non-PI, noncalcium mechanism.

IL-1-Induced Diacylglycerol in T Cells through Phosphatidylcholine (PC) Turnover

In recent studies by Rosoff et al. [135], similar findings were reported using the Jurkat cell line. As noted above, IL-1 amplifies the gene expression for mitogen or ionophore-primed Jurkat cells [130]. These cells produce increased IL-2 when stimulated by a combination of phorbol esters and IL-1 or anti-CD3 and IL-1. However, despite this activation, IL-1 does not increase cytosolic calcium in these cells or in neutrophils [70]. On the other hand, IL-1 causes a rapid rise in diacylgylcerol production. If this were due to hydrolysis of PI, generation of inositol triphosphate would have occurred. But similar to the results of others [134], this did not occur. These results suggest that IL-1 induces phospholipid hydrolysis through a non-PI turnover pathway, for example PC.

When the Jurkat cells were labeled with ^{3}H-choline and then stimulated with IL-1, phosphorylcholine increased dramatically (within 5 s) in the supernatant whereas the concentration of PC decreased in the organic phase of the cells. The increase in supernatant phosphorylcholine was not due to PC turnover via protein kinase C-stimulated diacylglycerol PI turnover because that pathway leads to increased cytosolic PC hydrolysis and the IL-1-stimulated PC turnover is primarily observed in the extracellular compartment. The concentration of IL-1 which induced the PC turnover was 30 f*M*, reaching a maximum at 100 f*M*, the concentration of IL-1 that stimulates T-cell proliferation and other biological effects in vitro. The IL-1-induced PC turnover was also observed in the D10.G4.1 T-helper cell line, the EL4 cell line and in nylon wool purified human peripheral blood T cells.

When the Jurkat, EL4 and D10.G4.1 cells were examined for IL-1 receptor affinity and numbers, we observed that Jurkat cells have no high affinity (see below) IL-1 receptors whereas the EL4 and D10.G4.1 cells studied with the same labeled IL-1α and under the same equilibrium conditions clearly contained the typical high affinity (5–50 p*M*) 80-kd IL-1 receptor. These studies suggest that the high affinity receptors for IL-1 are probably unrelated to the initiation of PC turnover and that this may be due to activation of a second, very low affinity receptor. This may explain the discrepancies in the biological response of many cells to IL-1 and the number of high affinity receptors.

Regardless of the proposed existence of a very low IL-1 receptor-IL-1 binding being the mechanism by which IL-1 stimulates PC turnover, the fact that there are at least two pathways by which diacylglycerol becomes elevated in T cells may explain the comitogen effect of IL-1 in a variety of immunocompetent cells (and even some nonimmunocompetent cells). T cells are rich in protein kinases and several isoenzymes with different tissue distribution exist. It has been proposed that the different protein kinases differ in the binding sites of the diacylglycerols (DAG) and that this may be determined by the type of fatty acid side chain. For example, mitogen which stimulates DAG through a PI turnover pathway results in activation of one type of protein kinase C, whereas IL-1 which stimulates DAG through PC turnover stimulates another isoenzyme. In some cells like the Jurkat, there is a requirement for both (?synergism) protein kinases before IL-2 gene expression is initiated. The mitogen (in this case anti-CD3) is not sufficient without phorbol esters or IL-1 to provide the second signal, in this case another protein kinase through a PC-specific DAG. This may be one explanation for the amplifying effect of IL-1 in a variety of immunocompetent cells. In comparing the response of EL4 cells to IL-1, ionophore and phorbol esters, Truneh et al. [136] concluded that IL-1 and TPA induce protein kinase C activation by two different mechanisms. It is thus probable that these two mechanisms reflect the two different pathways by which IL-1 and mitogens stimulate DAG.

Effect of IL-1 on Peripheral Blood T Cells

Employing human peripheral blood T cells activated by immobilized antibodies directed against the CD3 cell surface protein complex, soluble IL-1 serves as a cofactor for activation in the absence of macrophages [137]. Under these conditions, the addition of soluble recombinant IL-1 partially restores the role of the macrophage and one observes increased synthesis of IL-2 and expression of the IL-2 receptor. Increased DNA synthesis followed these events as an indicator of cell proliferation. However, when they are diluted, the ability of soluble IL-1 to replace the macrophage function is lost, suggesting that total macrophage depletion was not accomplished. IL-2 will, on the other hand, drive these cells to proliferate.

Effect of IL-1 on B-Cell Activation

Many investigators have shown that B cells and other cells serve as accessory cells in antigen recognition but have failed to demonstrate a role for IL-1 because IL-1 cannot be detected in B-cell supernatants. This issue may have been resolved by studies which demonstrate that B cells produce IL-1 [138, 139] and that B cells express membrane-bound IL-1 [19]. In fact, nearly all cells which can act as accessory cells produce IL-1. These include astrocytes, mesangial cells, keratinocytes and endothelial cells. Effects of IL-1 on B cells and immunoglobulin production continue to be made [140, 141]. The function of IL-1 in B-cell activation seems to be similar to that shown in T cells, that is, IL-1 acts as a helper or cofactor during the activation process particularly together with IL-4, also known as B-cell-stimulating factor-1. IL-1 activates B cells and contributes to formation of antibody. The first experiments to show the critical role for IL-1 in B-cell activation were performed with anti-IL-1 [142]. Adding antibody early to human peripheral blood mononuclear cells stimulated with pokeweed mitogen completely prevented B-cell activation and subsequent antibody formation. Other studies demonstrated that IL-1 synergized with various B-cell growth and differentiation factors leading to increased proliferation and antibody formation [143, 144]. Some of the biological activity of the natural B-cell growth and differentiation factors may have been due to the presence of IL-6 (see below). IL-1's ability to synergize with IL-6 on T-cell activation probably extends to B-cell activation. IL-1's ability to induce other B-cell-stimulating factors, including IL-6, interferon-γ and IL-2 must also be considered since these substances activate B cells.

The recent evidence demonstrating that IL-1 induces the production of B-cell-stimulating factor-2/hybridoma growth factor should be considered as a role for IL-1 in B-cell and immunoglobulin synthesis. The hybridoma growth factor protein also had the ability to stimulate plasmacytoma cell growth. The N-terminus of the plasmacytoma/hybridoma growth factor matches that of an interferon-β_2 [114, 115]. Others have also reported a B-cell-stimulating factor (BSF-2) [145] which is the same sequence as the 26-kd interferon-β_2. Interferon-β_2 has also been identified as the factor previously known as hepatocyte-stimulating factor, which caused an increase in CRP as well as other hepatic acute phase proteins [146]. Interferon-β_2 with its activity on B cells and liver cells is now named IL-6. It seems that IL-1's role in augmenting B-cell function and ultimately leading to the production of protective antibodies may be through its ability to induce the synthesis of B-cell-stimulating factors and/or up-regulating their receptors.

IL-1 and NK Cells

IL-1 also plays a role in host defense against tumor cells. There is evidence that IL-1 increases the binding of NK cells to tumor targets and that tumor cells induce synthesis of IL-1 by NK cells [147]. In addition, NK cells from patients with large tumor burdens produce significantly less IL-1 and have decreased killing ability than cells from healthy individuals. When incubated with exogenous IL-1, impaired NK function is restored. IL-1 also augments the binding of NK cells from healthy donors to tumor targets. Since IL-1 induces interferon and since interferon synergizes with IL-1 with respect to its actions on NK cells [148], one could view both mechanisms as an efficient aspect of host defense against tumors. However, unlike augmentation of T and B cell responses by IL-1 which are enhanced at febrile temperatures [149, 150], the effect of IL-1 on NK cells is reduced by febrile temperatures [151].

NK cells also produce IL-1 [152]. Recent studies suggest that in addition to endotoxin, high dose IL-2 is a stimulant for IL-1 and TNF production by NK cells [153]. The ability of IL-2 to induce IL-1 and TNF production appears to be due to the CD4 receptor on monocytes. This is up-regulated by γ-interferon which also increases the responsiveness of NK cells to IL-1-induced cytokine production.

The IL-1 Receptor

The initial studies on the binding of radiolabeled IL-1 were carried out using a variety of cells [154]; nearly all subsequent studies have focused on two cell types: murine thymoma cells and fibroblasts [reviewed in 155]. From these early studies, there appeared to be a single class of intermediate affinity receptors (kd ranged from 400 p*M* to 1 n*M*) and relatively few receptors (200/cell). Subsequent studies showed that the receptor number and affinity could be higher on thymoma cell lines and fibroblasts. Although the IL-1 receptor was specific in that it did not recognize other cytokines, the IL-1 receptor did not distinguish between IL-1α or IL-1β [67, 98, 132, 133, 156, 157]. In general, the binding correlated with the capacity of the cells to respond to IL-1. However, it soon became clear that there were two classes of IL-1 receptors [97, 132]. There is little question that an IL-1 receptor has been recognized by all groups and has a variable dissociation constant of 200–600 p*M*. From cross-linking experiments, this binding protein is probably what is often called the IL-1 receptor: an 80-kd glycosylated peptide whose unglycosylated form is approximately 65 kd. This molecule has been solubilized [158] and cloned; the sequence reveals that the IL-1 receptor belongs to the immunoglobulin superfamily [159].

The cloned molecule likely represents the high affinity receptor described by Lowenthal and MacDonald [132] and Bird and Saklatvala [97]. These groups described binding sites with an affinity of 5–10 p*M*, a considerably higher affinity than that previously reported. Furthermore, it has also been reported that both high and low affinity receptors exist on a wide variety of T-cell lines and normal cells. We have also observed two classes of IL-1 receptors on the D10.G4.1 murine T-helper cell line. In the case of our own studies [127], receptors of both high and low affinity were detected simultaneously on a subclone (D10S) of these cells. The failure to detect both classes of receptors in some studies may be due to the low specific activity of the radiolabeled IL-1 or that the low affinity receptor is not easily detectable.

In addition to the conflicting data on the IL-1 receptor affinity, there are also low numbers of receptors. However, Lowenthal and MacDonald [132] and Savage et al. [127] have reported much higher numbers of IL-1 receptors on murine EL4 cells and murine T-helper cells, respectively, but the high numbers (about 20,000/cell) are not in the same range as those observed for other growth factor receptors which can be of the order of 100,000–200,000/cell.

Comparison of the IL-1 and IL-2 Receptor

In evaluating the significance of high and low affinity receptors for IL-1 on different cells, some investigators have compared the situation with that of the receptors for IL-2. IL-2 receptors also exist as two classes distinguished by their ligand-binding affinities [160, 161]. The IL-2 receptor is made up of two distinct polypeptide chains (the so-called 'TAC' 55-kd antigen and a lower affinity polypeptide chain of 75 kd), each of which may bind to IL-2. One of the components binds with a lower affinity than the other; the two may both interact and combine with IL-2 to produce the high affinity receptor. The binding of the IL-2 to its 75-kd intermediate affinity receptor is sufficient to activate Na^+/H^+ exchange in YT 202 [162], a cell line expressing only the 75-kd IL-2-binding component. The same situation has been proposed for the IL-1 receptor [163]. Our own studies in D10S cells suggest that the IL-1 receptor comprises more than one polypeptide chain [127] and others have also observed the existence of a second chain [163].

Studies on the IL-2 receptor have revealed that high and low affinity receptors differ functionally and that a similar case can be made for the two classes of IL-1 receptors [132]. It has been proposed that the high affinity IL-1 receptor is incorporated into the cell and the number of IL-1 molecules endocytosed by various T-cell lines correlates with the number of high affinity IL-1 receptors expressed by these cells. This has been demonstrated for IL-2, where ligand internalization is believed to be mediated only by high affinity receptors [164]. In contrast, the internalization of IL-1 appears to be via a receptor with an intermediate affinity of 150 p*M* [37]. Internalization of IL-1 by a large granular lymphocyte cell line was demonstrated by receptors with a kd of about 100 p*M* [156, 157], which is higher than the high affinity receptors of 5–10 p*M*. One explanation for this discrepancy is that two different cell types have been studied, the fibroblast and the T cell, respectively.

Both IL-1α and IL-1β Bind to the Same Receptor

Most studies have demonstrated that using either form of IL-1 (α or β) in receptor-binding experiments, the one IL-1 form effectively competes with the binding of the other form. However, some studies suggest that although human and murine IL-1α may bind to human endothelial cells with an equal affinity, there is an unequal ability to induce a biological response [165]. Recombinant murine IL-1α was found to be 250- to 1,250-fold less active than recombinant human IL-1α in inducing endothelial cell adherence of human lymphocytes. Furthermore, Bird et al. [163] have shown that the concentration of porcine IL-1β required to elicit half maximal IL-2 production from NOB-1, a subline of murine thymoma IL-4, was 100-fold greater than that for porcine IL-1α. However, both forms of IL-1 were equally active at similar doses when acting on BALB/C 3T3 fibroblasts to increase lactate production.

These results and those of other investigators suggest that, contrary to what was first believed, the receptor-binding site on IL-1α and IL-1β may not be recognizing the same receptor loci in the different cells. Some attempts have been made to localize the receptor-binding site or specific recognition peptides on IL-1. MacDonald et al. [36] produced various mutations of the histidine residue at position 147 in IL-1β and showed that this resulted in up to a 100-fold reduction in receptor-binding affinity. Short synthetic peptide fragments of human IL-1β with immunostimulatory activity [25, 26] have been produced. These peptides code for hydrophobic regions of the IL-1β molecule which share significant amino acid homologies to those regions in the IL-1α molecule. Although they are biologically active, large amounts are required. Furthermore, there are no data suggesting that these peptides block the binding of full-length IL-1β to cells. Recently, attention has focused on the carboxyl terminal of the IL-1β structure. X-ray crystallographic studies of IL-1β have revealed that the histidine at position 147 as well as the

N-terminus and carboxyl terminus are exposed and available for membrane interaction [34]. In one study, antibodies produced to the C-terminal (amino acids 247–269) blocked IL-1 biological activities whereas antibodies produced to the N-terminal amino acids (117–134) had no effect [39]. On the other hand, mutations in the IL-1β N-terminus have dramatically affected IL-1 binding [37]. It appears that both the N- and C-termini interact with the receptor.

Regulation of IL-1 Receptor Expression

Both up-regulation and down-regulation of the IL-1 receptor may occur. Up-regulation of the IL-1 receptor on T cells has been reported on human peripheral T cells or murine splenic T cells stimulated with ConA [166, 167] and human mononuclear cells treated with corticosteroids increase receptor number. EL4 cells treated with transretinoic acid increase their binding of IL-1 [67]. Down-regulation of the IL-1 receptor has been reported by IL-1 itself [127, 156, 168].

Physical Structure of the IL-1 Receptor

Cross-linking experiments in a variety of cells identify a major binding protein of 80 kd on SDS-PAGE. A second cross-linked species has been observed at 116 kd. It has been suggested that the higher molecular mass species could either represent a protein of 116 kd or a tertiary complex of the 80-kd receptor, a 25-kd binding protein and IL-1. Cross-linking studies in rat brain indicate that the rat brain IL-1 receptor also has a molecular mass of 80 kd [169]. It may be significant that these authors also found a second IL-1-binding protein in rat brain at 58 kd. The homogeneous IL-1 receptor, purified from EL4 6.1 cells, has been found to be a protein of 80 kd which correlates with the results found for the affinity cross-linking experiments.

Most experimental evidence to date has confirmed the existence of a plasma membrane protein of molecular size ranging from 70 to 80 kd for the IL-1 receptor. The Raji B-lymphoma cells have a lower binding affinity but much higher receptor density (kd = 2.1 n*M*, 7,709 sites/cell) [37] than the murine T cells (kd = 400 p*M*, 241 sites/cell). Cross-linking studies showed that the IL-1 receptor in the B cells had a lower molecular mass than that in the T cells (68 kd compared to 80 kd). We, however, have found that the D10S subclone of the murine T-helper cell D10.G4.1 possesses a high number of IL-1 receptors which would imply that the cells do not differ from the B cells in this regard [127]. It is likely that variations in size of the IL-1-binding components in the different cells are caused by glycosylation. In fact, treatment of EL4 as well as fibroblast-derived IL-1 receptors with glycanases reduces the molecule size to about 50 kd [Bird et al., pers. commun.].

In our laboratory we have digested radioactive IL-1 cross-linked proteins from EL4 and D10S cells with glycanase and generated bands 10–15 kd lower in mass [Savage et al., unpubl. observations]. Most authors have also reported on the presence of minor radioactive bands either of higher or lower molecular mass than the major species. We have observed a 25-kd IL-1-binding protein in D10S cells which was not observed with EL4 cells. It is possible that the 116-kd protein is a combination of the 80- and 25-kd proteins. The significance of these lower molecular weight IL-1-binding components is not clear; however, a recent report on the structure of the IL-1 receptor may explain the molecular weight differences (see below).

Alternatively, random lateral association is occurring with neighboring membrane proteins or additional proteins may be involved in the signal transduction mechanism, both leading to cross-linked IL-1 proteins with various molecular weights. Again, as in the case of the IL-2 receptor, there may be more than one polypeptide comprising the IL-1 receptor complex. Indeed, there exists some evidence to support the latter possibility [163]. A recent report depicts the molecular structure of the IL-1 receptor belonging to the immunoglobulin class [159]. In fact, it appears that there is a repeating sequence of light-chained immunoglobulin molecules with molecular weights of about 12.5 kd. This would implicate IL-1 receptors as being polymeric. It would explain the lower molecular weight 25-kd IL-1-binding protein (low affinity) as being a dimer and that the deglycosylated form of 50 kd to represent further molecular polymerization. The higher forms of IL-1-binding proteins would also fit into this scheme. Further experimentation will clarify this issue. Of similar importance is the likelihood that the IL-1 receptor, being part of the immunoglobulin class, would be rearranged on some cells. The significance is unclear. However, one of the overriding aspects of the IL-1 receptor is that cells respond to subpicomolar concentrations of IL-1 without possessing demonstrable IL-1 binding. One explanation is that these cells have very low affinity receptors and the receptor occupancy is slow, short and of such low affinity that binding cannot be easily demonstrated. However, binding presumably takes place with sufficient interaction that signal transduction takes place.

Postreceptor-Binding Events

Information concerning the intracellular events following the IL-1-1/IL-1 receptor complex has only recently been studied. No clear picture has emerged. Of importance when considering the postreceptor events of IL-1 action is the fact that in addition to the growth-promoting properties of IL-1, the cytokine has a wide variety of other biologic activities. These properties can only be explained if there are different receptors on the various cells, or alternatively the receptors are identical but the postreceptor machinery differ in the various cells in which the biological events occur. When considering the cellular proliferating effects, some evidence suggests that, like other growth factors such as EGF and insulin, IL-1 may activate a tyrosine kinase, although it is not known whether the kinase is part of the IL-1 receptor. However, the same authors have also observed plasma membrane protein phosphorylation as a result of IL-1 action on membranes from K562 cells. Surprisingly the K562 cells have been shown to possess very small numbers of receptors (<10 molecules bound per cell) and the effect of IL-1 in these cells is not to produce proliferation, but rather induce killing mechanisms. Do the two phosphorylation events induced by IL-1 differ? Results from our laboratory [G. von Bulow et al., unpubl.] show that recombinant IL-1 acts on the plasma membrane isolated from EL4 cells can also induce the phosphorylation of certain membrane proteins, although the site of phosphorylation is unknown at present. On the other hand, Matsushima et al. [170] have observed the phosphorylation of a cytosolic 65-kd protein when IL-1 acts on normal human peripheral blood mononuclear leukocytes pretreated with glucocorticoids. Phosphorylation was observed in serine residues of the protein and not on tyrosine residue, as described by Martin et al. [171] for K562 cells.

In addition to the phosphorylation effects of IL-1, internalization of the ligand occurs. Internalization of IL-1 has been shown by several groups [97, 156, 157, 168]. Some studies show subsequent lysosomal trafficking of ligand and reutilization of receptors. After 3 h at 37 °C, the ligand was located in the lysosomal fraction and there was an increase in the TCA-soluble fraction at 6 h. In addition, some studies show electron microscopic evidence, using autoradiographic detection techniques, for the appearance of radioactive IL-1 within the nucleus of the cells. There was little evidence for the degradation of IL-1 by the cells, at least up to 4–6 h. Lowenthal and MacDonald [172] have also observed the uptake of IL-1 in EL4 cells and that a significant fraction of internalized IL-1 was found in the nucleus.

Another postreceptor event that has been described for IL-1 is the ion flux of both Na^+ and Ca^{++} across the plasma membrane of a murine pre-B-cell line [173]. Major questions require clarification of the nature of the signal initiated by the formation of the cell surface IL-1 complex and the role of internalization of the IL-1 ligand in the transfer of information. The finding that IL-1 will rapidly increase PC turnover in Jurkat cells which do not manifest a detectable IL-1 receptor underscores the complexity of IL-1 effects on cells and the postexposure events. It is possible that there are both receptor (high affinity) events leading to PI turnover and the generation of DAG in some cells and low affinity receptor interaction resulting in the generation of another DAG from PC turnover. Moreover, both systems may work in some cells. The end result would perhaps provide a number of activated protein kinases.

Comparison of IL-1, TNF and IL-6

IL-1 and TNF

TNF was initially identified in the circulation of animals following the injection of endotoxin. It was also discovered in the supernates from stimulated macrophage cell lines where its property as an inhibitor of lipoprotein lipase led to its being named 'cachectin' since it produced a wasting syndrome when chronically administered to mice. The amino acid sequences of TNF [174] are identical to cachectin [175]. TNF, a product of stimulated monocytes and macrophages, is also produced by lymphocytes [176], endothelial cells and keratinocytes. A structurally related polypeptide, initially isolated from activated T cells, is lymphotoxin. Lymphotoxin and TNF produce similar biological changes in a variety of cells. The amino sequence of TNF and lymphotoxin are closely related [174] and both molecules are recognized by the same cell membrane receptor. Like IL-1β and α, TNF and lymphotoxin are sufficiently structurally distinct molecules that antibodies produced to each cytokine do not cross-react with the other cytokine. Although originally studied for its ability to kill tumor cells in vitro as well as when injected into tumor-bearing mice, the widespread biological effects of TNF on mesenchymal and other cells have been the focus of studies related to its inflammatory properties, particularly in mediating synovial cell activity and cartilage and bone degradation. Moreover, recombinant human TNF has been injected into human subjects and many of its systemic effects such as fever, leukopenia and hypotension, which were studied in animals, have now been observed in humans [177].

The biological properties of TNF share remarkable similarities to those of IL-1, particularly the nonimmunological effects of IL-1. Some lymphocyte-activating properties of IL-1 or IL-6 are shared with TNF but these require considerably higher concentrations of TNF than IL-1. There are recent reports on the ability of TNF to activate T cells [178] including the expression of IL-2 receptors [179]. B cells are also stimulated by TNF [68, 180]. Compared to IL-1, the molar concentration of TNF required to stimulate immunocompetent cells is one or two orders of magnitude greater than IL-1 [68, 178]. Since TNF induces the synthesis and release of immunostimulatory polypeptides such as IL-1 and IL-6 from monocytes, fibroblasts, and endothelial cells, it is possible that these cytokines augment the action of TNF on lymphocytes. Some investigators have attempted to separate a direct action of TNF on lymphocytes from that secondary to the induction of IL-1 or IL-6 [178]. It also appears that unlike IL-1, the immunostimulatory effects of TNF are species specific.

Nearly every nonimmunological biological property of IL-1 has also been observed with TNF. These include fever [181], the induction of PGE_2 and collagenase synthesis in a variety of tissues [91], bone and cartilage resorption [182], inhibition of lipoprotein lipase [14], increases in hepatic acute phase proteins, complement components, and a decrease in albumin synthesis [80]. Slow wave sleep and appetite suppression are also observed following the injection of TNF [183]. As discussed above, both molecules induce fibroblast proliferation collagen synthesis [100]. The cytotoxic activity of TNF differs from that of IL-1 in that IL-1 is inactive on a variety of tumor targets for which TNF is a potent cytotoxin. However, IL-1 exhibits cytotoxic effects on melanoma cells which are unaffected by TNF [184]. Another difference between IL-1 and TNF is that IL-1 can function as a cofactor for stem cell activation (hemopoietin-1 activity) [10, 11, 185], whereas TNF suppresses bone marrow colony formation [186]. Both IL-1 and TNF induce the synthesis of colony-stimulating factors [112, 113].

Similar to IL-1, TNF induces fever by its direct ability to stimulate hypothalamic PGE_2 synthesis [181]. Levels of circulating TNF rise rapidly in human subjects injected with endotoxin [187, 188] and are associated with the symptoms of the prodrome and chill period of the fever. In addition to fever, TNF produces hypotension, leukopenia and local tissue necrosis [106]. On a weight basis in rabbits, TNF is more potent than IL-1 in producing a shock [69]. Administration of anti-TNF antibodies to rabbits prevents the shock induced by endotoxin [189]. The shock-like responses to TNF likely reflect the effects on the vascular endothelium. TNF stimulates PGI_2, PGE_2 and platelet-activating factor production by cultured endothelium. In addition, like IL-1, TNF stimulates procoagulant activity, leukocyte adherence and plasminogen activator inhibitor on these cells. TNF also induces a

Table 3. Comparison of biological properties of IL-1 and TNF

Biological property	IL-1	TNF
Endogenous pyrogen fever	+	+
Slow wave sleep	+	+
Hemodynamic shock	+	+
Increased hepatic acute phase protein synthesis	+	+
Decreased albumin synthesis	+	+
Activation of endothelium	+	+
Decreased lipoprotein lipase	+	+
Decreased cytochrome P-450	+	+
Decreased plasma Fe/Zn	+	+
Increased fibroblast proliferation	+	+
Increased synovial cell collagenase and PGE_2	+	+
Induction of IL-1	+	+
T/B cell activation	+	±
Hemopoietin-1 activity	−	+

capillary leak syndrome. Despite the similarities, receptors for TNF and IL-1 are distinct and specific, and receptor binding to the respective ligand is only displaced by the specific cytokine. Furthermore, IL-1 down-regulates its own receptor as well as that of TNF. The most likely explanation is that TNF and IL-1 stimulate similar intracellular messages by different pathways and alter the same cascade of intracellular metabolites. Table 3 lists the biological similarities between IL-1 and TNF. Of note is the fact that TNF stimulates human neutrophil oxidative metabolism [190] whereas IL-1 does not [70]; on the other hand, IL-1 induces histamine and arylsulfatase release from human basophils [75] and eosinophils, respectively [77], and TNF does not. There is another macrophage product with a molecular weight of 8 kd which is chemotactic for neutrophils and is clearly not IL-1 nor TNF [191, 192]. This factor may have been present in some preparations of natural IL-1 and TNF and could have accounted for their effects on neutrophils. The biological properties of this macrophage product in comparison to IL-1 and TNF remain to be ascertained.

Synergism between IL-1 and TNF

The effects of IL-1 or TNF on a variety of cells in vitro as well as systemic effects in vivo are often biologically indistinguishable. When the two cytokines are used together in experimental studies, the net effect often exceeds the additive effect of each cytokine. Potentiation or frank synergism between these two molecules has been demonstrated in several studies. On fibroblasts, IL-1 and TNF act synergistically in the production of PGE_2 [193]. The cytotoxic effect of TNF and IL-1 on certain tumor cells is also synergistic in vitro and, when administered to tumor-bearing mice in vivo, both molecules act synergistically to eliminate the tumor [M. Palladino, pers. commun.]. IL-1 combined with TNF protects rats exposed to lethal hyperoxia [194]. IL-1 and TNF act synergistically in the induction of radioprotection [195, 196]. IL-1 induces cytotoxic effects on the insulin-producing β cells of the islets of Langerhans; this effect is dramatically augmented by TNF [197, 198]. IL-1 and TNF induce neutrophilic infiltration when injected intradermally into experimental animals; when injected together, these cytokines act synergistically and can replace endotoxin in the generation of the local Shwartzman reaction [86]. Rats receiving intravenous infusions of IL-1 or TNF manifest metabolic changes reflected in plasma amino acid levels, but when given together, negative nitrogen balance and muscle proteolysis can be demonstrated [199]. Although high doses (10–20 μg/kg) of TNF produce a shock-like state with tissue damage [106], IL-1 and TNF act synergistically to produce hemodynamic shock and pulmonary hemorrhage at doses of only 1 μg/kg when given together [69]. The two cytokines also act synergistically in the aggregation of neutrophils and the synthesis of thromboxanes [89]. Considering the fact that IL-1 and TNF are often present together in human body fluids, including inflammatory joint fluid, the synergism between the two cytokines cannot be considered a laboratory observation. The synergism between these two cytokines seems to be due to second message molecules rather than up-regulation of cell receptors; in fact, IL-1 reduces TNF receptors [200].

IL-1 and IL-6

The recent cloning and expression of B-cell-stimulating factor-2 (B-cell growth factor-2) [145] and hybridoma growth factor [201] has supported the observation that these molecules are identical to interferon-β_2 [114, 115]. Interferon-β_2 was cloned and expressed before its identification as B-cell-stimulating factor-2. In addition, interferon-β_2 appears to be the same molecule previously termed 'hepatocyte-stimulating factor' [146], inducing a variety of hepatic acute phase proteins in cultured liver cells. The molecule interferon-β_2 hybridoma growth factor, plasmacytoma growth factor, B-cell-stimulating factor-2 and hepatocyte -stimulating factor is now termed IL-6. Like IL-1 and TNF, IL-6 is an endogenous pyrogen and an inducer of acute phase responses. In a clinical study, serum levels of IL-6 correlated with the amount of fever present in patients with burn injuries [202]. IL-6 levels have also been reported elevated in patients undergoing renal rejection [203] and in the cerebrospinal fluid of patients with central nervous system infections [204]. Although IL-6 is an inducer of fibrinogen synthesis in hepatic cell lines, these cultured cells require the presence of corticosteroids to observe a response [205]. In mice and rats, IL-6 does not induce fibrinogen unless corticosteroids are administered at the same time [206]; IL-1, on the other hand, induces large amounts of fibrinogen without the requirement of such cofactors [207].

IL-6's effects on lymphocytes and bone marrow stem cells are broadly based. Attention has focused on the ability of natural or recombinant IL-6 to act as a 'lymphocyte-activating factor' in the typical murine thymocyte comitogenesis assay. Similar to recombinant IL-1a or IL-1b, recombinant IL-6, in the presence of ConA, induces the production of IL-2 from cytotoxic T-cell lines (CTLL) [208]. A similar response was observed when recombinant IL-6 was added with mitogen or antibody to the T-cell antigen receptor-stimulated peripheral blood CD4$^+$ T-cell depleted of accessory cells. IL-2 was released and acted as the proliferative signal for these cells. Using high concentrations (100 units) of recombinant IL-1, production of IL-2 was not observed [208], leading these authors to the conclusion that IL-6 rather than IL-1 was providing the 'first signal' along with mitogen or anti-T-cell receptor activation in mature circulating T cells. The molar concentration of the recombinant IL-6 used in these studies is difficult to ascertain and limiting dilutions of the purified T-cell targets were not carried out. Therefore, like IL-1, it is difficult to ascertain whether IL-6 activates lymphocytes in the absence of macrophages, particularly since IL-6 is a product of activated macrophages; moreover, hemopoietic growth factors [209], as well as IL-6 are inducible by IL-1 [114, 115, 210].

IL-6 will act as a comitogen for human thymocytes and macrophage-depleted human T cells [211]. However, unlike IL-1, the comitogenesis effect of IL-6 was not inhibited by antibodies to the IL-2 receptor [211]. Similar studies have been reported for thymic and peripheral blood T cells of the L3T4$^+$ and Lyt-2$^+$ subsets [212]. Recombinant IL-6 acts as an autocrine growth factor for human myeloma cells in vitro [213] and IL-1 has also been shown to act as an autocrine growth factor for EBV-transformed human B cells [214]. In addition, recombinant IL-6 induces the production of immunoglobulin from activated B cells in the absence of growth [215].

We have compared recombinant IL-1 and IL-6 in a variety of assays. In the murine thymocyte comitogenesis assay, human recombinant IL-1

Table 4. Comparison of IL-1, TNF and IL-6

Biological property	IL-1	TNF	IL-6
Endogenous pyrogen fever	+	+	+
Hepatic acute phase proteins	+	+	+
T-cell activation	+	±	+
B-cell activation	+	±	+
B-cell Ig synthesis	±	–	+
Fibroblast proliferation	+	+	–
Stem cell activation (hemopoietin-1)	+	–	+
Nonspecific resistance to infection	+	+	+
Radioprotection	+	+	±
Synovial cell activation	+	+	–
Endothelial cell activation	+	+	–
Induction of IL-1 and TNF from monocytes	+	+	–
Induction of IL-6	+	+	–

stimulates proliferation at 1–10 pg/ml whereas 1–10 ng/ml of recombinant IL-6 is required for the same amount of proliferation. This may be due to species specificity of IL-6. Using a highly IL-1-sensitive murine T-helper cell line (D10.G4.1), 5–6 orders of magnitude greater concentrations of IL-6 compared to IL-1 are required for a proliferative signal. In rabbits, IL-6 behaves as an endogenous pyrogen, producing a rapid-onset monophasic fever; however, 20- to 50-fold greater amounts of IL-6 are required to produce the same elevation in body temperature as that following IL-1 [216]. Once again, this may be due to species specificity. However, when recombinant *human* IL-1 was compared to recombinant human IL-6 on *human* monocyte PGE_2 production, 50- to 100-fold more IL-6 than IL-1 was required. Similar dose-response differences have been observed using human synovial cell or fibroblasts as targets [Dayer, pers. commun.]. Unlike IL-1 and TNF, IL-6 does not induce IL-1 or TNF; in fact, IL-6 suppresses endotoxin- and TNF-induced IL-1 production [216]. IL-6 does not activate endothelial cells in vitro [217, 218]. In general, IL-6 appears to be a weak inflammatory peptide. Of considerable importance is the observation that IL-1 and IL-6 both act as hemopoietin-1 on bone marrow cultures [10, 11, 219]. In addition, IL-6 protects granulocytopenic mice against lethal gram-negative infection [van der Meer, pers. commun.] similar to the protection afforded by IL-1 [220]. IL-1 and IL-6 act synergistically in protecting mice given lethal irradiation [R. Neta, pers. commun.]. The lack of inflammatory properties and positive effects on B- and T-cell functions as well as bone marrow and nonspecific host-defense mechanisms make IL-6 potentially useful in treating some diseases, especially bone marrow transplantation. Table 4 lists the biological activities of IL-1, TNF and IL-6.

Acknowledgments

These studies are supported by NIH Grant AI15614. The author thanks Professor Nerina Savage for her contribution to the section on IL-1 receptors. The author also acknowledges the suggestions and help of the following persons: H.G. Cannon, B. Clark, S. Endres, R. Ghorbani, T. Ikejima, G. Lonnemann, J.W. Mier, L. Miller, R. Numerof, S.F. Orencole, R. Schindler, S.D. Sisson, J.W.M. van der Meer, A. Vanstory, S.J.C. Warner, and S.M. Wolff.

References

1 Atkins E: The pathogenesis of fever. Physiol Rev 1960;40:580–646.
2 Murphy PA, Chesney J, Wood WB Jr: Further purification of rabbit leukocyte pyrogen. J Lab Clin Med 1974;83:310–322.
3 Dinarello CA, Renfer L, Wolff SM; Human leukocytic pyrogen; purification and development of a radioimmunoassay. Proc Natl Acad Sci USA 1977;74:4624–4627.
4 Kampschmidt RF: Leukocytic endogenous mediator/endogenous pyrogen; in Powanda, Canonico (eds): Physiologic and Metabolic Responses of the Host. Amsterdam, Elsevier/North-Holland, 1981, pp 55–74.
5 Dinarello CA: Interleukin-1. Rev Infect Dis 1984;6:51–95.
6 Gery I, Waksman BH: Potentiation of the T-lymphocyte response to mitogens. II. The cellular source of potentiating mediator(s). J Exp Med 1972;136:143–155.
7 Krane SM, Dayer J-M, Simon LS, et al: Mononuclear cell-conditioned medium containing mononuclear cell factor (MCF), homologous with interleukin-1, stimulates collagen and fibronectin synthesis by adherent rheumatoid synovial cells: effects of prostaglandin E_2 and indomethacin. Collagen Relat Res 1985;5:99–117.
8 Saklatvala J, Sarsfield SJ, Townsend Y: Pig interleukin-1. Purification of two immunologically different leukocyte proteins that cause cartilage resorption, lymphocyte activation, and fever. J Exp Med 1985;162:1208–1222.
9 Dewhirst FE, Stashenko PP, Mole JE, et al: Purification and partial sequence of human osteoclast-activating factor: identity with interleukin-1-beta. J Immunol 1985;135:2562–2568.
10 Moore MA, Warren DJ: Synergy of interleukin-1 and granulocyte colony-stimulating factor: in vivo stimulation of stem-cell recovery and hematopoietic regeneration following 5-fluorouracil treatment of mice. Proc Natl Acad Sci USA 1987;84:7134–7137.

11 Mochizuki DY, Eisenman JR, Conlon PJ, et al: Interleukin-1 regulates hematopoietic activity, a role previously ascribed to hemopoietin. Proc Natl Acad Sci USA 1987;84:5267–5271.
12 Thomas KA, Rios-Candelore M, Gimez-Gallego, et al: Pure brain-derived fibroblast growth factor is a potent angiogenic vascular endothelial cell mitogen with sequence homology to interleukin-1. Proc Natl Acad Sci USA 1985;82:6409–6413.
13 Dinarello CA: Interleukin-1: amino acid sequences, multiple biological activities and comparison with tumor necrosis factor (cachectin). Year Immunol 1986;2:68–89.
14 Beutler B, Cerami A: Recombinant interleukin-1 suppresses lipoprotein lipase activity in 3T3-L1 cells. J Immunol 1985;135:3969–3971.
15 Dayer J-M, Beutler B, Cerami A: Cachectin/tumor necrosis factor stimulates collagenase and prostaglandin E_2 production by human synovial cells and dermal fibroblasts. J Exp Med 1985;162:2163–2168.
16 Auron PE, Webb AC, Rosenwasser LJ, et al: Nucleotide sequence of human monocyte interleukin-1 precursor cDNA. Proc Natl Acad Sci USA 1984;81:7907–7911.
17 Lomedico PT, Gubler U, Hellman CP, et al: Cloning and expression of murine interleukin-1 in *Escherichia coli*. Nature 1984;312:458–462.
18 Lepe-Zuniga B, Gery I: Production of intracellular and extracellular interleukin-1 (IL-1) by human monocytes. Clin Immunol Immunopathol 1984;31:222–230.
19 Kurt-Jones EA, Kiely JM, Unanue ER: Conditions required for expression of membrane IL-1 on B-cells. J Immunol 1985;135:1548–1550.
20 Clark BD, Collins KL, Gandy MS, et al: Genomic sequence for human prointerleukin-1-beta: possible evolution from a reverse transcribed prointerleukin-1-alpha gene. Nucl Acids Res 1986;14:7897–7905.
21 Furutani Y, Notake M, Fuki T, et al: Complete nucleotide sequence of the gene for human interleukin-1-alpha. Nucleic Acids Res 1986;14:3167–3179.
22 Webb AC, Collins KL, Auron PE, et al: Interleukin-1 gene (IL-1) assigned to long arm of human chromosome 2. Lymphokine Res 1986;5:77–85.
23aRimsky L, Wakasugi H, Ferrara P, et al: Purification to homogeneity and NH2-terminal amino acid sequence of a novel interleukin-1 species derived from a human B cell line. J Immunol 1986;136:3304–3310.
23bTagaya Y, Okada M, Sugie K, et al: IL-2 receptor (p55)/TAC-inducing factor: Purification and characterization of adult T cell leukemia-derived factor. J Immunol 1988;140:1–7.
24 Auron PE, Rosenwasser LJ, Matsushima K, et al: Human and murine interleukin-1 possess sequence and structural similarities. J Mol Immunol 1985;2:169–177.
25 Antoni H, Presentini R, Perin F, et al: A short synthetic peptide fragment of human interleukin-1 with immunostimulatory but not inflammatory activity. J Immunol 1986;137:3201–3204.
26 Rosenwasser LJ, Webb AC, Clark BD, et al: Expression of biologically active human interleukin-1 subpeptides by transfected simian COS cells. Proc Natl Acad Sci USA 1986;83:1–4.
27 Massone A, Baldari C, Censini SD, et al: Mapping of biologically relevant sites on human IL-1-beta using monoclonal antibodies. J Immunol 1988;140:3812–3816.
28 Cannon JG, Dinarello CA: Increased plasma interleukin-1 activity in women after ovulation. Science 1985;227:1247–1249.

29 Kimball ES, Pickeral SF, Oppenheim JJ, et al: Interleukin-1 activity in normal human urine. J Immunol 1984;133:256–260.
30 Dinarello CA, Clowes GHA Jr, Gordon AH, et al: Cleavage of human interleukin-1: isolation of a peptide fragment from plasma of febrile humans and activated monocytes. J Immunol 1984;133:1332–1338.
31 Palaszynski EW: Synthetic C-terminal peptide of IL-1 functions as a binding domain as well as an antagonist for the IL-1 receptor. Biochem Biophys Res Commun 1987; 147:204–211.
32 Auron PE, Warner SJC, Webb AC, et al: Studies on the molecular nature of human interleukin-1. J Immunol 1987;138:1447–1456.
33 Pierart ME, Najdovski T, Appelboom TE, et al: Effect of human endopeptidase 24.11 ('enkephalinase') on IL-1-induced thymocyte proliferation activity. J Immunol 1988;140:3808–3811.
34 Priestle JP, Schaer H-P, Gruetter MG: Crystal structure of the cytokine interleukin-1-beta. Eur Mol Biol Org J 1988;7:339–343.
35 Cohen FE, Dinarello CA: Structural homology between interleukin-1 and tumor necrosis factor. J Leuk Biol 1987;42:548.
36 MacDonald HR, Wingfield P, Schmeissner U, et al: Point mutations of human interleukin-1 with decreased receptor binding affinity. FEBS Lett 1986;209:295–298.
37 Horuk R, Huang JJ, Covington M, et al: A biochemical and kinetic analysis of the interleukin-1 receptor. J Biol Chem 1987;262:16275–16278.
38 Dinarello CA, Cannon JG, Mier JW, et al: Multiple biological activities of human recombinant interleukin-1. J Clin Invest 1986;77:1734–1739.
39 Riveau G, Dinarello CA, Chedid L: Inhibition of human interleukin-1 induced T-cell proliferative response by an antisynthetic peptide antiserum. J Immunol, in press.
40 Kern JA, Lamb RJ, Reed JC, et al: Dexamethasone inhibition of interleukin-1-beta production by human monocytes. Posttranscriptional mechanisms. J Clin Invest 1988;81:237–244.
41 Schindler R, Dinarello CA: Regulation of IL-1-beta in RNA and protein synthesis in human mononuclear cells. Lymphokine Res 1988;7:273.
42 Fenton MJ, Clark BD, Collins KL, et al: Transcriptional regulation of the human prointerleukin-1-beta gene. J Immunol 1987;138:3972–3979.
43 Fenton MJ, Vermeulen MW, Clark BD, et al: Human pro-IL-1-beta gene expression in monocytic cells is regulated by two distinct pathways. J Immunol 1988;140:2267–2273.
44 Libby P, Ordovas JM, Auger KR, et al: Endotoxin and tumor necrosis factor induce interleukin-1 gene expression in adult human vascular endothelial cells. Am J Pathol 1986;124:179–186.
45 Libby P, Ordovas JM, Auger KR, et al: Inducible interleukin-1 gene expression in vascular smooth muscle cells. J Clin Invest 1986;78:1432–1438.
46 Dinarello CA, Ikejima T, Warner SJC, et al: Interleukin-1 induces interleukin-1. I. Induction of circulating interleukin-1 in rabbits in vivo and in human mononuclear cells in vitro. J Immunol 1987;139:1902–1910.
47 Warner SJC, Auger KR, Libby R: Interleukin-1 induces IL-1 gene expression in smooth muscle cells. J Exp Med 1987;165:1316–1331.
48 Knudsen PJ, Dinarello CA, Strom TB: Glucocorticoids inhibit transcriptional and post-transcriptional expression of interleukin-1 in U937 cells. J Immunol 1987; 139:4129–4134.

49 Kunkel SL, Chensue SW: Arachidonic acid metabolites regulate interleukin-1 production. Biochem Biophys Res Commun 1985;128:892–897.
50 Knudsen PJ, Dinarello CA, Strom TB: Prostaglandins post-transcriptionally inhibit monocyte expression of intracellular cyclic adenosine monophosphate. J Immunol 1986;137:3189–3194.
51 Matsushima K, Taguchi M, Kovacs EJ, et al: Intracellular localization of human monocyte associated interleukin-1 activity and release of biologically active IL-1 from monocytes by trypsin and plasmin. J Immunol 1986;136:2883–2891.
52 Lisi PJ, Chu C-W, Koch GA, et al: Development and use of a radioimmunoassay for human interleukin-1-beta. Lymphokine Res 1987;6:229–244.
53 Endres S, Ghorbani R, Lonnemann G, et al: Measurement of immunoreactive interleukin-1-beta from human mononuclear cells: optimization of recovery, intrasubject consistency and comparison with interleukin-1-alpha and tumor necrosis factor. Clin Immunol Immunopathol 1988;49:424–438.
54 Demczuk S, Baumberger C, Mach D, et al: Expression of human IL-1-alpha and beta messenger RNA's and IL-1 activity in human peripheral blood mononuclear cells. J Mol Cell Immunol 1987;3:255–258.
55 Endres S, Cannon JG, Ghorbani R, et al: In vitro production IL-1b, IL-1a, tumor necrosis factor and IL-2 in a large cohort of human subjects: distribution, effect of cyclooxygenase inhibition and evidence of independent gene regulation. J Immunol, in press.
56 Endres S, Ghorbani R, Kelley VE, et al: The effect of dietary supplementation with N-3 polyunsaturated fatty acids on the synthesis of interleukin-1 and tumor necrosis factor. N Engl J Med, in press, 1989.
57 Ghezzi P, Dinarello CA: Interleukin-1 induces interleukin-1. III. Specific inhibition of interleukin-1 production by gamma interferon. J Immunol, in press.
58 Rola-Pleszczynski M, Lemaire L: Leukotrienes augment interleukin-1 production by human monocytes. J Immunol 1985;135:3958–3961.
59 Dinarello CA, Bishai I, Rosenwasser LJ, et al: The influence of lipoxygenase inhibitors in the in vitro production of human leukocytic pyrogen and lymphocyte activating factor (interleukin-1). Int J Immunopharmacol 1984;6:43–50.
60 Singer II, Scott S, Hall GL, et al: Interleukin-1-beta is localized in the cytoplasmic ground substance but is largely absent from the Golgi apparatus and plasma membranes of stimulated human monocytes. J Exp Med 1988;167:389–407.
61 Sisson SD, Dinarello CA: Recombinant granulocyte-macrophage colony-stimulating factor stimulates interleukin-1-beta, interleukin-1-alpha and tumor necrosis factor production. Blood 1988;72:1368–1374.
62 Folks TM, Justement J, Kinter A, et al: Cytokine-induced expression of HIV-1 in a chronically infected promonocyte cell line. Science 1987;238:800–802.
63 Bakouche O, Brown DC, Lachman LB: Subcellular localization of human monocyte interleukin-1: evidence for an inactive precursor molecule and a possible mechanism for IL-1 release. J Immunol 1987;138:4249–4253.
64 Beuscher HU, Fallon RJ, Colten HR: Macrophage membrane interleukin-1 regulates the expression of acute phase proteins in human hepatoma Hep 3B cells. J Immunol 1987;139:1896–1901.
65 Kobayashi Y, Appella E, Yamada M, et al: Phosphorylation of intracellular precursors of human IL-1. J Immunol 1988;140:2279–2287.

66 Van Damme J, De Ley M, Opdenakker G, et al: Homogeneous interferon-inducing 22K factor is related to endogenous pyrogen and interleukin-1. Nature 1985; 314:266–268.

67 Kilian PL, Kaffka KL, Stern AS, et al: Interleukin-1-alpha and interleukin-1-beta bind to the same receptor on T-cells. J Immunol 1986;136:4509–4514.

68 Kehrl JH, Miller A, Fauci AS: Effect of tumor necrosis factor alpha on mitogen-activated human B cells. J Exp Med 1987;166:786–791.

69 Okusawa S, Gelfand JA, Ikejima T, et al: Interleukin-1 induces a shock-like state in rabbits: Synergism with tumor necrosis factor and the effect of cyclooxygenase inhibition. J Clin Invest 1988;81:1162–1172.

70 Georgilis K, Schaefer C, Dinarello CA, et al: Human recombinant interleukin-1-beta has no effect on intracellular calcium or on functional responses of human neutrophils. J Immunol 1987;138:3403–3407.

71 Goldberg AL, Kettelhut IC, Furuno K, et al: Activation of protein breakdown and prostaglandin E_2 production in rat skeletal muscle in fever is signaled by a macrophage product distinct from interleukin-1 or other known monokines. J Clin Invest 1988;81:1378–1383.

72 Moldawer LL, Svaninger G, Gelin J, et al: Interleukin-1 and tumor necrosis factor do not regulate protein balance in skeletal muscle. Am J Physiol 1987;253:C773–C779.

73 Clowes GHA Jr, George BC, Bosari S, et al: Induction of muscle protein degradation by recombinant interleukin-1 (rIL-1) and its spontaneously occurring fragments. J Leuk Biol 1987;42:547.

74 Rhyne JA, Mizel SB, Wheeler JG, et al: Action of interleukin-1 on interleukin-1 receptor bearing polymorphonuclear leukocytes. Clin Immunol Immunopathol 1988;43:354–361.

75 Subramanian N, Bray MA: Interleukin-1 releases histamine from human basophils and mast cells in vitro. J Immunol 1987;138:271–277.

76 Haak-Frendscho M, Dinarello CA, Kaplan AP: Recombinant human interleukin-1-beta causes histamine release from human basophils. J Allergy Clin Immunol, 1988;82:218–223.

77 Pincus SH, Whitcomb EA, Dinarello CA: Interaction of interleukin-1 and TPA in modulation of eosinophil function. J Immunol 1986;137:3509–3514.

78 Ramadori G, Sipe JD, Dinarello CA, et al: Pretranslational modulation of acute phase hepatic protein synthesis by murine recombinant interleukin-1 and purified human IL-1. J Exp Med 1985;162:930–942.

79 Ghezzi P, Saccardo B, Villa P, et al: Role of interleukin-1 in the depression of liver drug metabolism by endotoxin. Infect Immun 1986;54:837–840.

80 Perlmutter DH, Dinarello CA, Punsal P, et al: Cachectin/tumor necrosis factor regulates hepatic acute phase gene expression. J Clin Invest 1986;78:1349–1354.

81 Perlmutter D, Goldberger G, Dinarello CA, et al: Regulation of class III major histocompatability complex gene products by interleukin-1. Science 1986;232:850–852.

82 Rossi V, Breviario F, Ghezzi P, et al: Interleukin-1 induces prostacyclin in vascular cells. Science 1985;229:1174–1176.

83 Dejana E, Breviario F, Erroi A, et al: Modulation of endothelial cell function by different molecular species of interleukin-1. Blood 1987;69:695–699.

84 Bevilacqua MP, Pober JS, Wheeler ME, et al: Interleukin-1 acts on cultured human vascular endothelial cells to increase the adhesion of polymorphonuclear leukocytes, monocytes and related leukocyte cell lines. J Clin Invest 1985;76:2003–2011.

85 Dustin ML, Rothelin R, Bhan AK, et al: Induction by interleukin-1 and interferon-gamma, tissue distribution, biochemistry and function of a natural adherence molecule (ICAM-1). J Immunol 1986;137:245–254.
86 Movat HZ, Burrowes CE, Cybulsky MI, et al: Acute inflammation and a Shwartzman-like reaction induced by interleukin-1 and tumor necrosis factor: Synergistic action of the cytokines in the induction of inflammation and microvascular injury. Am J Pathol 1987;129:463–467.
87 Bevilacqua MP, Pober JS, Majeau GR, et al: Interleukin-1 induces biosynthesis and cell surface expression of procoagulant activity on human vascular endothelial cells. J Exp Med 1984;160:618–623.
88 Nachman RL, Hajjar KA, Silverstein RL, et al: Interleukin-1 induces endothelial cell synthesis of plasminogen activator inhibitor. J Exp Med 1986;163:1545–1547.
89 Conti P, Cifone MG, Alesse E, et al: In vitro enhanced thromboxane B_2 release by polymorphonuclear leukocytes and macrophages after treatment with human recombinant interleukin-1. Prostaglandins 1986;32:111–115.
90 Giulian D, Woodward J, Young DG, et al: Interleukin-1 injected into mammalian brain stimulates astrogliosis and neovascularization. J Neurosci 1988;8:709–714.
91 Dayer J-M, de Rochemonteix B, Burrus B, et al: Human recombinant interleukin-1 stimulates collagenase and prostaglandin E_2 production by human synovial cells. J Clin Invest 1986;77:645–648.
92 Schnyder J, Payne T, Dinarello CA: Human monocyte or recombinant interleukin-1s are specific for the secretion of a metalloproteinase from chondrocytes. J Immunol 1987;138:496–503.
93 Pettipher ER, Higgs GA, Henderson B: Interleukin-1 induces leukcoyte infiltration and cartilage proteoglycan degradation in the synovial joint. Proc Natl Acad Sci USA 1986;83:8749–8753.
94 Phadke K: Fibroblast growth factor enhances the interleukin-1-mediated chondrocytic protease release. Biochem Biophys Res Commun 1987;142:448–452.
95 Schmidt JA, Oliver CN, Lepe-Zuniga JL, et al: Silica-stimulated monocytes release fibroblast proliferation factors identical to interleukin-1. J Clin Invest 1984; 73:1462–1468.
96 Canalis E: Interleukin-1 has independent effects on DNA and collagen synthesis in cultures of rat calvariae. Endocrinology 1986;118:74–81.
97 Bird TA, Saklatvala J: Identification of a common class of high affinity receptors for both types of porcine interleukin-1 on connective tissue cells. Nature 1986;324:263–265.
98 Chin J, Cameron PM, Rupp E, et al: Identification of a high-affinity receptor for native human interleukin-1-beta and interleukin-1-alpha on normal human lung fibroblasts. J Exp Med 1987;165:70–80.
99 Matsushima K, Bano M, Kidwell WR, et al: Interleukin-1 increases collagen type IV production by murine mammary epithelial cells. J Immunol 1985;134; 904–909.
100 Kohase M, May LT, Tamm I, et al: A cytokine network in human diploid fibroblasts: interactions of beta-interferons, tumor necrosis factor, platelet-derived growth factor and interleukin-1. Mol Cell Biol 1987;7:273–280.
101 Gimenez-Gallego G, Rodkey J, Bennett C, et al: Brain-derived acidic fibroblast growth factor: complete amino acid sequence and homologies. Science 1985; 230:1385–1388.

102 Shimokado K, Raines EW, Madtes DK, et al: A significant part of macrophage derived growth factor consists of at least two forms of PDGF. Cell 1985;43:277–283.
103 Martinet Y, Rom WN, Grotendorst GR, et al: Exaggerated spontaneous release of platelet-derived growth factor by alveolar macrophages from patients with idiopathic pulmonary fibrosis. N Engl J Med 1987;317:202–208.
104 Duel TF, Senior RM, Huang JS, et al: Chemotaxis of monocytes and neutrophils to platelet-derived growth factor. J Clin Invest 1982;69:1056–1062.
105 Tzeng DY, Duel TF, Huang JS, et al: Platelet-derived growth factor promotes polymorphonuclear leukocyte activation. Blood 1984;64:123–127.
106 Tracey KJ, Beutler B, Lowry SF, et al: Shock and tissue injury induced by recombinant human cachectin. Science 1986;234:470–473.
107 Scarborough DE, Dinarello CA, Reichlin S: Recombinant human interleukin-1-beta increases somatostatin in fetal rat hypothalamic neurons in vitro. J Leuk Biol 1987;42:560.
108 Roh MS, Drazenovich KA, Jeffrey BS, et al: Direct stimulation of the adrenal cortex by interleukin-1. Surgery 1987;102:140–146.
109 McCarthy DO, Kluger MJ, Vander AJ: Suppression of food intake during infection: Is interleukin-1 involved? Am J Clin Nutr 1987;42:1179–1187.
110 Bendtzen K, Mandrup-Poulsen T, Nerup J, et al: Human pI 7 interleukin-1 is cytotoxic for pancreatic islets of Langerhans. Science 1986;232:1545–1547.
111 Giulian D, Lachman LB: Interleukin-1 stimulation of astroglial proliferation after brain injury. Science 1985;228:497–500.
112 Zucali JR, Dinarello CA, Gross MA, et al: Interleukin-1 stimulates fibroblasts to produce granulocyte-macrophage colony-stimulating activity (GM-CSA) and prostaglandin E_2. J Clin Invest 1986;77:1857–1863.
113 Bagby GC Jr, Dinarello CA, Wallace P, et al: Interleukin-1 stimulates GM-CSF release by vascular endothelial cells. J Clin Invest 1986;78:1316–1323.
114 Van Damme J, De Ley M, Van Snick J, et al: The role of interferon-beta and the 26-kd protein (interferon-beta) as mediators of the antiviral effect of interleukin-1 and tumor necrosis factor. J Immunol 1987;139:1867–1872.
115 Van Damme J, Opdenakker G, Simpson RJ, et al: Identification of the human 26-kd protein, interferon-beta-2 (IFN-beta-2), as a B cell hybridoma/plasmacytoma growth factor induced by interleukin-1 and tumor necrosis factor. J Exp Med 1987;165:914–919.
116 Breder CD, Dinarello CA, Saper CB: Interleukin-1 immunoreactive innervation of the human hypothalamus. Science 1988;240:321–324.
117 Ho S, Abraham RT, Nilson A, et al: Interleukin-1-mediated activation of interleukin-4 producing T lymphocytes. Proliferation by IL-4-dependent and IL-4-independent mechanisms. J Immunol 1987;139:1532–1537.
118 Tartakovsky B, Kovacs EJ, Takacs L, et al: T cell clone producing an IL-1-like activity after stimulation by antigen-presenting B cells. J Immunol 1986;137:160–165.
119 Smith KA, Lachman LB, Oppenheim JJ, et al: The functional relationship of the interleukins. J Exp Med 1980;151:1551–1561.
120 Simic MM, Stosic-Grujicic S: The dual role of interleukin-1 in lectin-induced proliferation of T cells. Folia biol 1985;31:410–424.
121 Maennel DN, Mizel SB, Diamantstein T, et al: Induction of interleukin-2 responsiveness in thymocytes by synergistic action of interleukin-1 and interleukin-2. J Immunol 1985;134:3108–3110.

122 Loegdberg L, Wassmer P, Shevach EM: Role of the L3T4 antigen in T-cell activation. Cell Immunol 1985;94:299–311.
123 Loegdberg L, Shevach EM: Role of the Ly-1 antigen in interleukin-1-induced thymocyte activation. Eur J Immunol 1985;15:1007–1013.
124 Kaye J, Gillis S, Mizel SB, et al: Growth of a cloned helper T-cell line induced by a monoclonal antibody specific for the antigen receptor: interleukin-1 is required for the expression of receptors for interleukin-2. J Immunol 1984;133:1339–1345.
125 Orencole SF, Ikejima T, Cannon JG, et al: A subclone of D10.G4.1 T-cells which specifically proliferates in response to interleukin-1 at attograms/ml in the absence of mitogen (abstract). Lymphokine Res 1987;6:1210.
126 Lacey DL, Chappel JC, Teitelbaum SL: Interleukin-1 stimulates proliferation of a nontransformed T lymphocyte line in the absence of co-mitogen. J Immunol 1987; 139:2649–2655.
127 Savage N, Orencole SF, Dinarello CA: Interleukin-1 induced T-lymphocyte proliferation. II. Demonstration of molecularly distinct IL-1 receptor on a subclone of cells and its differential modulation by IL-1, IL-2 and IL-4. J Immunol, in press.
128 Kupper T, Horowitz M, Lee F, et al: Autocrine growth of T cells independent of interleukin-2: identification of interleukin-4 (IL-4, BSF-1) as an autocrine growth factor for a cloned antigen-specific helper T cell. J Immunol 1987;138: 4280–4284.
129 Kupper T, Flood P, Coleman D, et al: Growth of an interleukin-2/interleukin-4-dependent T cell line induced by granulocyte-macrophage colony-stimulating factor (GM-CSF). J Immunol 1987;138:4288–4292.
130 Arya SK, Gallo RC: Transcriptional modulation of human T-cell growth factor gene by phorbol ester and interleukin-1. Biochemistry 1984;23:6685–6690.
131 Simon PL: Calcium mediates one of the signals required for interleukin-1 and 2 production by murine cell lines. Cell Immunol 1984;87:720–726.
132 Lowenthal JW, MacDonald HR: Binding and internalization of interleukin-1 by T cells. Direct evidence for high- and low-affinity classes of interleukin-1 receptor. J Exp Med 1986;164:1060–1074.
133 Lowenthal JW, Cerottini JC, MacDonald HR: Interleukin-1-dependent induction of both interleukin-2 secretion and interleukin-2 receptor expression by thymoma cells. J Immunol 1986;137:1226–1231.
134 Abraham RT, Ho SN, Barna TJ, et al: Transmembrane signaling during interleukin-1-dependent T cell activation. J Biol Chem 1987;262:2719–2728.
135 Rosoff PM, Savage N, Dinarello CA: Interleukin-1 stimulates diacylglycerol production in T lymphocytes by a novel mechanism. Cell 1988;54:73–81.
136 Truneh A, Simon P, Schmitt-Verhulst AM: Interleukin-1 and protein kinase C activator are dissimilar in their effects on IL-2 receptor expression and IL-2 secretion by T lymphocytes. Cell Immunol 1986;103:365–374.
137 Williams JM, DeLoria D, Hansen JA, et al: The events of primary T cell activation can be staged by use of Sepharose-bound anti-T3 (64.1) monoclonal antibody and purified interleukin-1. J Immunol 1985;135:2249–2255.
138 Matsushima K, Procopio A, Abe H, et al: Production of interleukin-1 activity by normal human peripheral blood B-cells. J Immunol 1985;135:1132–1136.
139 Scala G, Kuang YD, Hall RE, et al: Accessory cell function of human B cells. Production of both interleukin-1-like activity and an interleukin-1 inhibitory factor by an EBV-transformed human B cell line. J Exp Med 1984;159:1637–1652.

140 Muraguchi A, Kehrl JH, Butler JL, et al: Regulation of human B-cell activation, proliferation, and differentiation by soluble factors. J Clin Immunol 1984;4:337–347.
141 Lipsky PE: Role of interleukin-1 in human B-cell activation. Contemp Top Mol Immunol 1985;10:195–217.
142 Lipsky PE, Thompson PA, Rosenwasser LJ, et al: The role of interleukin-1 in human B cell activation: inhibition of B cell proliferation and the generation of immunoglobulin secreting cells by an antibody against human leukocytic pyrogen. J Immunol 1983;130:2708–2714.
143 Falkoff RJM, Muraguchi A, Hong JX, et al: The effect of interleukin-1 on human B cell activation and proliferation. J Immunol 1983;131:801–805.
144 Falkoff RJM, Butler JL, Dinarello CA, et al: Direct effects of a monoclonal B cell differentiation factor and of purified interleukin-1 on B cell differentiation. J Immunol 1984;133:692–696.
145 Hirano T, Yasukawa K, Harada H, et al: Complementary DNA for a novel human interleukin (BSF-2) that induces B lymphocytes to produce immunoglobulin. Nature 1986;324:73–76.
146 Gauldie J, Richards C, Harnish D, et al: Interferon-beta-2/B-cell stimulatory factor type 2 shares identity with monocyte-derived hepatocyte-stimulating factor and regulates the major acute phase protein response in liver cells. Proc Natl Acad Sci USA 1987;84:7251–7256.
147 Herman J, Dinarello CA, Kew MC, et al: The role of interleukin-1 in tumor NK cell interactions. Correction of defective NK cell activity in cancer patients by treating target cells with IL-1. J Immunol 1985;135:2882–2886.
148 Dempsey RA, Dinarello CA, Mier JW, et al: The differential effects of human leukocytic pyrogen/lymphocyte activating factor, T-cell growth factor and interferon on human natural killer cells. J Immunol 1982;129:2504–2510.
149 Duff GW, Durum SK: Fever and immunoregulation: hyperthermia, interleukins-1 and -2 and T-cell proliferation. Yale J Biol Med 1982;55:437–442.
150 Hanson DF, Murphy PA: Temperature sensitivity of interleukin-dependent murine T-cell proliferation: Q2 mapping of the responses to peanut agglutinin-negative thymocytes. J Immunol 1985;135:3011–3020.
151 Dinarello CA, Dempsey RA, Allegretta M, et al: Inhibitory effects of elevated temperature on cytokine production and natural killer activity. Cancer Res 1986; 46:6235–6241.
152 Scala G, Allaven P, Djew JX, et al: Human large granular lymphocytes are potent producers of interleukin-1. Nature 1984;309:56–59.
153 Numerof RP, Dinarello CA, Endres S, et al: Interleukin-2 stimulates the production of interleukin-1-beta, interleukin-1-alpha, and tumor necrosis factor-alpha from human peripheral blood mononuclear cells. J Immunol, in press, 1989.
154 Dower SK, Kronheim SR, March CJ, et al: Detection and characterization of high affinity plasma membrane receptors for human interleukin-1. J Exp Med 1985; 162:501–515.
155 Savage N, Dinarello CA; Interleukin-1 and its receptor. Crit Rev Immunol, in press.
156 Matsushima K, Yodoi J, Tagaya Y, et al: Down-regulation of interleukin-1 receptor expression by IL-1 and fate of internalized ^{125}I-labeled IL-1-beta in a human large granular lymphocyte cell line. J Immunol 1986;137:3183–3188.

157 Matsushima K, Akahoshi T, Yamada M, et al: Properties of a specific interleukin-1 receptor on human Epstein-Barr virus transformed B-lymphocytes: Identity of the receptor for IL-1-alpha and 1 beta. J Immunol 1986;136:4496–4502.
158 Paganelli KA, Stern AS, Kilian PL: Solubilization of the interleukin-1 receptor. J Immunol 1987;138:2249–2253.
159 Sims JE, March CT, Cosman DJ, et al: The receptor for interleukin-1 is a member of the immunoglobulin superfamily. Science 1988;241:585–588.
160 Lowenthal JW, Zubler RH, Nabholz M, et al: Similarities between interleukin-2 receptor number affinity on activated B and T lymphocytes. Nature 1985;315:675–678.
161 Robb RJ, Greene WC, Rusk CM: Low and high cellular receptors for interleukin-2. Implications for binding of the Tac antigen. J Exp Med 1986;160:1126–1136.
162 Mills GB, May C: Binding of interleukin-2 to 75-kd intermediate affinity receptor is sufficient to trigger Na^+/H^+ exchange. J Immunol 1987;139:4083–4088.
163 Bird TA, Gearing AJH, Saklatvala J: Murine interleukin-1 receptor: differences in binding properties between fibroblastic and thymoma cells and evidence for a two chain receptor model. FEBS Lett 1987;225:21–26.
164 Weissman AM, Hartford JB, Svetlik PB, et al: Only high affinity receptors of intermediate internalization of ligand. Proc Natl Acad Sci USA 1986;83:1463–1467.
165 Thieme TR, Hefeneider SH, Wagner CR, et al: Recombinant murine and human IL-1-alpha bind to human endothelial cells with an equal affinity, but have an unequal ability to induce endothelial cell adherence of lymphocytes. J Immunol 1987;139:1173–1178.
166 Shirakawa F, Tanaka Y, Ota T, et al: Expression of interleukin-1 receptors on peripheral T cells. J Immunol 1987;138:4243–4249.
167 Dower SK, Urdal DL: The interleukin-1 receptor. Immunol Today 1987;8:46–48.
168 Mizel SB, Kilian PL, Lewis JC, et al: The interleukin-1 receptor: Interleukin-1 binding and internalization in T cell and fibroblasts. J Immunol 1987;138:2906–2912.
169 Farrar WL, Kilian PL, Ruff MR, et al: Visualization and characterization of interleukin-1 in brain. J Immunol 1987;139:459–463.
170 Matsushima K, Kobayashi Y, Copeland TD, et al: Phosphorylation of a cytosolic 65-dalton protein induced by interleukin-1 in glycocorticoid pretreated human peripheral blood mononuclear leukocytes. J Immunol 1987;139:3367–3372.
171 Martin M, Lovett DH, Resch K: Interleukin-1 specific phosphorylation of a 41-kd plasma membrane in the human tumor cell line K562. Immunobiol. 1986;171:145–150.
172 Lowenthal JW, MacDonald HR: Expression of interleukin-1 receptor is restricted to the $L3T4^+$ subset of mature T lymphocytes. J Immunol 1987;138:1–3.
173 Stanton TH, Maynard M, Bomsztyk K: Interleukin-1 on intracellular concentration of sodium calcium and potassium in 70Z/3 cells. J Biol Chem 1986;2:701–705.
174 Pennica D, Nedwin GE, Hayflick JS, et al: Human tumor necrosis factor: precursor structure, expression and homology to lymphotoxin. Nature 1984;312:724–729.
175 Beutler B, Cerami A: Cachectin and tumor necrosis factor: two sides of the same biological coin. Nature 1986;320:584–588.
176 Cuturi MC, Murphy M, Costa-Gomi MP, et al: Independent regulation of tumor necrosis factor and lymphotoxin production by human peripheral blood lymphocytes. J Exp Med 1987;165:1581–1594.
177 Chapman PB, Lester TJ, Casper ES, et al: Clinical pharmacology of recombinant human tumor necrosis factor in patients with advanced cancer. J Clin Oncol 1987; 5:1942–1951.

178 Ranges GE, Zlotnik A, Espevik T, et al: Tumor necrosis factor alpha/cachectin is a growth factor for thymocytes. J Exp Med 1988;167:1472–1478.
179 Plaetinck G, Declercq W, Tavernier J, et al: Recombinant tumor necrosis factor can induce interleukin-2 receptor expression and cytolytic activity in a rat × mouse T cell hybrid. Eur J Immunol 1987;17:1835–1838.
180 Jelinek DF, Lipsky PE: Enhancement of human B cell proliferation and differentiation by tumor necrosis factor alpha and interleukin-1. J Immunol 1987;139:2970–2976.
181 Dinarello CA, Cannon JG, Wolff SM, et al: Tumor necrosis factor (cachectin) is an endogenous pyrogen and induces interleukin-1. J Exp Med 1986;163: 1433–1450.
182 Saklatvala J: Tumor necrosis factor-alpha stimulates resorption and inhibits synthesis of proteoglycan in cartilage. Nature 1986;322:547–550.
183 Shoham S, Davenne D, Cady AB, et al: Recombinant tumor necrosis factor and interleukin-1 enhance slow-wave sleep. Am J Physiol 1987;253:R142–R149.
184 Onozaki K, Matsushima K, Aggarwal BB, et al: Human interleukin-1 is a cytocidal factor for several tumor cell lines. J Immunol 1985;135:3962–3967.
185 Stanley ER, Bartocci A, Patinkin D, et al: Regulation of very primitive, multipotent, hemopoietic cells by hemopoietin-1. Cell 1986;45:667–673.
186 Zucali JR, Broxmeyer HE, Gross MA, et al: Recombinant human tumor necrosis factor alpha and beta stimulate fibroblasts to produce hemopoietic growth factors in vitro. J Immunol 1988;140:840–844.
187 Hesse DG, Tracey KJ, Fong Y, et al: Cytokine appearance in human endotoxemia and primate bacteremia. Surg Gynecol Obstet 1988;166:147–153.
188 Michie HR, Manogue KR, Spriggs DR, et al: Detection of circulating tumor necrosis factor during endotoxemia in humans. N Engl J Med 1988;318:1481–1486.
189 Mathison JC, Wolfson E, Ulevitch RJ: Participation of tumor necrosis factor/ cachectin in the mediation of gram-negative bacterial lipopolysaccharide induced injury in rabbits. J Clin Invest 1988;81:1925–1937.
190 Klebanoff SJ, Vadas MA, Harlan JM, et al: Stimulation of neutrophils by tumor necrosis factor. J Immunol 1986;137:2695–2699.
191 Wolpe SD, Davatelis G, Sherry B, et al: Macrophages secrete a novel heparin-binding protein with inflammatory and neutrophil chemokinetic properties. J Exp Med 1988;167:570–581.
192 Yoshimura T, Matsushima K, Tanaka S, et al: Purification of a human monocyte-derived neutrophil chemotactic factor that has peptide sequence similarity to other host defense cytokines. Proc Natl Acad Sci USA 1987;84:9233–9237.
193 Elias JA, Gustilo K, Baeder W, et al: Synergistic stimulation of fibroblast prostaglandin production by recombinant interleukin-1 and tumor necrosis factor. J Immunol 1987;138:3812–3816.
194 White CW, Ghezzi P, Dinarello CA, et al: Recombinant tumor necrosis factor/ cachectin and interleukin-1 pretreatment decreases lung oxidized glutathione accumulation, lung injury and mortality in rats exposed to hyperoxia. J Clin Invest 1987; 79:1863–1873.
195 Neta R, Douches S, Oppenheim JJ: Interleukin-1 is a radioprotector. J Immunol 1987;136:2483–2485.
196 Neta R, Oppenheim JJ, Douches SD: Interdependence of the radioprotective effects of human recombinant interleukin-1-alpha, tumor necrosis factor alpha, granulocyte

colony-stimulating factor, and murine recombinant granulocyte-macrophage colony-stimulating factor. J Immunol 1988;140:108–111.
197 Mandrup-Poulsen T, Bendtzen K, Dinarello CA, et al: Potentiation of IL-1 mediated B-cell killing by TNF. Human tumor necrosis factor potentiates human interleukin-1 mediated rat pancreatic B-cell cytotoxicity. J Immunol 1987;139:4077–4082.
198 Mandrup-Poulsen T, Bendtzen K, Nerup J, et al.: Affinity purified human interleukin-1 is cytotoxic to isolated islets of Langerhans. Diabetologia 1986;29:63–67.
199 Pomposelli JJ, Flores EA, Bistrian BR, et al: Dose response of recombinant mediators in the acute phase response. Clin Res 1987;35:514A.
200 Holtmann H, Wallach D: Down-regulation of the receptors for tumor necrosis factor by interleukin-1 and 4-beta-phorbol-12-myristate-13-acetate. J Immunol 1987;139: 1161–1177.
201 Brakenhoff JPJ, De Groot ER, Evers RF, et al: Molecular cloning and expression of hybridoma growth factor in *Escherichia coli*. J Immunol 1987;139:4116–4121.
202 Nijsten MWN, De Groot ER, Ten Duis HJ, et al: Serum levels of interleukin-6 and acute phase responses. Lancet 1987;ii;921.
203 van Oers MH, van der Heyden AA, Aarden LA: Interleukin-6 (IL-6) in serum and urine of renal transplant recipients. Clin Exp Immunol 1988;71:314–319.
204 Houssiau FA, Bukasa K, Sindic CJM, et al: Elevated levels of the 26K human hybridoma growth factor (interleukin-6) in cerebrospinal fluid of patients with acute infection of the central nervous system. Clin Exp Immunol 1988;71:320–323.
205 Andus T, Geiger T, Hirano T, et al: Recombinant human B cell stimulatory factor 2 (BSF-2/IFN-beta-2) regulates beta-fibrinogen and albumin mRNA levels in Fao-9 cells. FEBS Lett 1987;221:18–22.
206 Moshage HJ, Roelofs HM, van Pelt JF, et al: The effect of interleukin-1, interleukin-6 and its interrelationship on the synthesis of serum amyloid A and C-reactive protein in primary cultures of adult human hepatocytes. Biochem Biophys Res Commun 1988;155:112–117.
207 Bertini R, Bianchi M, Villa P, et al: Depression of liver drug metabolism and increased fibrinogen by interleukin-1 and tumor necrosis factor: a comparison with lymphotoxin and interferon. Int J Immunopharmacol, in press.
208 Garman RD, Jacobs KA, Clark SC, et al: B-cell-stimulatory factor 2 (beta-2-interferon) functions as a second signal for interleukin-2 production by mature murine T cells. Proc Natl Acad Sci USA 1987;84:7629–7633.
209 Yang YC, Tsai S, Wong GG, et al: Interleukin-1 regulation of hematopoietic growth factor production by human stromal fibroblasts. J Cell Physiol 1988;134:292–296.
210 Schindler R, Endres SD, Ghorbani R, et al: Comparison of immunoreactive IL-6 production with IL-1, IL-2 and tumor necrosis factor from the mononuclear cells of a large cohort of human subjects. Lymphokine Res, 1988;7:311.
211 Lotz M, Jirik F, Kabouridis P, et al: B cell stimulating factor 2/interleukin-6 is a costimulant for human thymocytes and T lymphocytes. J Exp Med 1988;167:1253–1258.
212 Uyttenhove C, Coulie PG, Van Snick J: T cell growth and differentiation induced by interleukin-HP1/IL-6, the murine hybridoma/plasmacytoma growth factor. J Exp Med 1988;167:1417–1427.
213 Kawano M, Hirano T, Matsuda T, et al: Autocrine generation and requirement of BSF-2/IL-6 for human multiple myelomas. Nature 1988;332:83–85.

214 Scala G, Morrone G, Tamburrini M, et al: Autocrine growth function of human interleukin-1 molecules on ROHA-9, an EBV-transformed human B cell line. J Immunol 1987;138:2527–2532.
215 Muraguchi A, Hirano T, Tang B, et al: The essential role of B cell stimulatory factor 2 (BSF-2/IL-6) for the terminal differentiation of B cells. J Exp Med 1988;167:332–344.
216 Dinarello CA, Cannon JG, Endres S, et al: Comparison of the multiple biological activities of recombinant IL-1 and IL-6. Lymphokine Res, in press.
217 Sironi M, Breviario F, Biondi A, et al: Interleukin-6 production in resting and interleukin-1 treated endothelial cells. J Immunol, in press.
218 Lapierre LA, Fiers W, Pober JS: Three distinct classes of regulatory cytokines control endothelial cell MHC antigen expression. Interactions with immune gamma interferon differentiate the effects of tumor necrosis factor and lymphotoxin from those of leukocyte alpha and fibroblast beta interferons. J Exp Med 1988;167:794–804.
219 Ikebuchi K, Wong GG, Clark SC, et al: Interleukin-6 enhancement of interleukin-3-dependent proliferation of multipotential hemopoietic progenitors. Proc Natl Acad Sci USA 1987;84:9035–9039.
220 van der Meer JWM, Barza M, Wolff SM, et al: Low dose recombinant interleukin-1 protects granulocytopenic mice from lethal gram-negative infection. Proc Natl Acad Sci USA 1988;85:1620–1623.

Charles A. Dinarello, MD, Division of Geographic Medicine and Infectious Diseases, New England Medical Center Hospitals, 750 Washington Street, Boston, MA 02111 (USA)

Sorg C (ed): Macrophage-Derived Cell Regulatory Factors.
Cytokines. Basel, Karger, 1989, vol 1, pp 155–172

Biosynthesis of Complement Components by Macrophages

Harry Martin, Michael Loos

Institute of Medical Microbiology of the Johannes Gutenberg University of Mainz, FRG

Introduction

Macrophages are within the first line of cellular immunity to invading macroorganisms. A wide variety of bacteria, yeasts and viruses are readily engulfed and destroyed by macrophages in the absence of antibody. This process occurs either where macrophages are strategically positioned to filter out foreign material, such as in the liver or lungs, or due to the macrophages' ability to migrate into an inflammatory site. Both the directed movement of a macrophage towards a region of inflammation (chemotaxis), and the engulfment of foreign matter which occurs on arrival of the macrophage in the region (phagocytosis), are heavily dependent on the function of the complement system, a family of 19 serum proteins. Activated complement proteins bound to the pathogen surface (opsonization) greatly enhance phagocytosis by virtue of complement receptors on the macrophage membrane. Macrophages themselves synthesize the majority of complement components. However, the bulk synthesis of serum proteins is carried out in the liver. Fibroblasts and epithelial cells are also known to produce complement proteins. In addition to contributing to the serum complement pool, macrophage complement biosynthesis serves as a complement source in the extravascular space where the complement concentration is low and where any activation is liable to result in rapid depletion of complement. In this regard it is noteworthy that about 50% of C3 in inflamed joints is synthesized locally [1].

The main purpose of this review is to summarize the biosynthesis of complement components by macrophages and the regulation of complement biosynthesis. However, to understand the biological function of the complement proteins produced by macrophages it seems to us worthwhile giving a brief overview of the complement system and to discuss the biological functions of the individual components.

The Complement System

Similar to other humoral systems, such as coagulation, fibronolysis and the kinin system, the complement (C) system is a multifactorial system whose activation generates a cascade of biologically active products which normally are beneficial but can, under certain circumstances, cause damage to the host. The 19 distinct serum proteins which comprise the complement system have been highly purified, physiochemically characterized and the genes of some of these proteins have been sequenced [for further details, see 2–8].

The complement system can be divided into three compartments: (a) the classical, (b) the alternative, pathways of activation, and (c) the terminal sequence, triggered by the classical or alternative pathways, which causes membrane damage and cell lysis. C3 occupies a central position in the complement system, being at the nexus of these three compartments. The classical and alternative pathways can be seen as mechanisms for generating C3 cleaving enzymes and each pathway has its control proteins which normally ensure that activation occurs only where appropriate.

The Enzymes of the Classical Pathway

The classical pathway [9] can be activated by antigen-antibody complexes and involves the C components C1, C4 and C2 (fig. 1). C1 by itself is a macromolecule composed of C1q and doublets of the proenzymes C1r and C1s which are linked by Ca^{++} ions ($C1r_2$-$C1s_2$). Only the antibodies IgM and IgG have a binding site for the C1q subcomponent of macromolecular C1. In addition to IgM and IgG, a variety of other substances interact directly with C1 and C1q [10], e.g.: bacterial-derived lipopolysaccharides (LPS) and lipid A, polyanionic substances, the C-reactive protein (CRP); envelopes of some RNA viruses, etc. Several steps are involved in the activation of C1 by IC: Binding of C1 to IC via C1q, the recognition unit of C1, induces a conformational change within the C1q molecule. This causes a further conformational change in C1r exposing the enzymatic site of one C1r

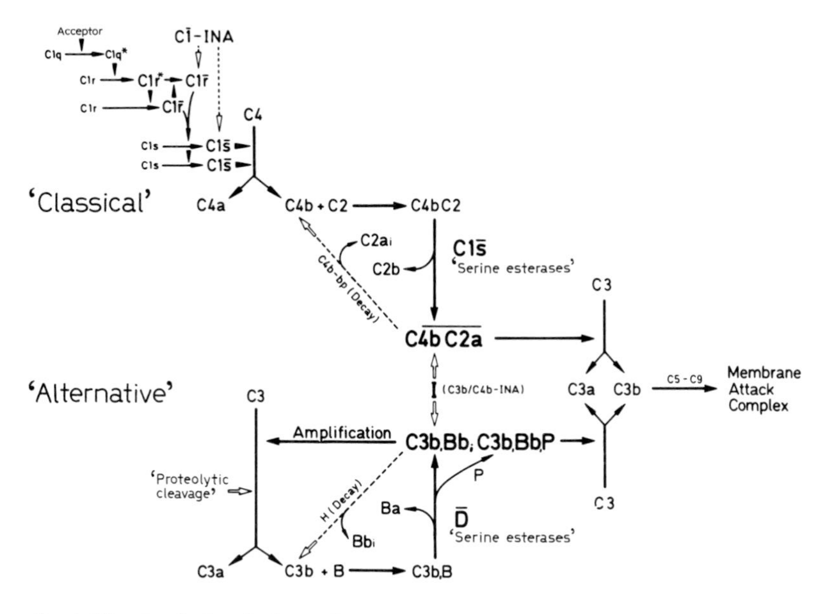

Fig. 1. The classical and alternative pathways of the complement system.

monomer which cleaves the second monomer of C1r to $C1\bar{r}$. $C1\bar{r}$ then cleaves C1s to $C1\bar{s}$ and is converting C1 to $C\bar{1}$. $C1\bar{r}$ and $C1\bar{s}$ are serine esterases. Both precursor forms are single chain molecules (85 kd) which are cleaved upon activation into a larger 'a' (56 kd) and a smaller 'b' (27 kd) chain which are linked by disulfide bonds. The two subcomponents differ in their substrate specificities: $C1\bar{r}$ can activate only C1s, but not C4 and C2; $C1\bar{s}$ cannot activate C1r, but it does cleave C4 and C2.

The natural substrates of $C1\bar{s}$ in serum are C4 and C2, the next two components in the C cascade of the classical pathway. Cleavage of C4 and C2 on cell surfaces by $C1\bar{s}$ leads to the formation of the classical pathway C3 convertase, C4b2a [11]. The component C4 is composed of three polypeptide chains, α, β, γ, with molecular weights of 90, 70, and 33 kd. When subjected to the action of $C1\bar{s}$, a 10-kd peptide (C4a) is cleaved from the α-chain and on the larger fragment of C4 (C4b) a highly labile binding site is exposed. Only when the cleavage of C4 takes place in the presence of acceptor sites for C4b, such as IC-$C\bar{1}$, are C4b sites generated, IC-$C\bar{1}$4b, for the further participation of C4 in the cascade. In the absence of acceptor sites for C4b the labile binding site on C4b is no longer available, and C4 becomes iC4b which is incapable of furthering the C cascade.

The formation of a hemolytically active C4b site provides the binding site for C2, the second natural substrate of C1$\bar{s}$. The generation of the classical C3 convertase (CV4b2a), is at least a two-step reaction. First, binding of native C2 to bound C4b, i.e. IC-C$\bar{1}$4b, is dependent on the presence of Mg^{++} ions. The conversion of the C4b2 complex into the C2 convertase is dependent on the enzymatic cleavage of C2 into C2b and C2a by C1$\bar{s}$. The larger fragment C2a remains bound to C4b and the formed C4b2a complex now becomes capable of cleaving the next component of the cascade, C3. The newly formed C4b2a is no longer dependent upon Mg^{++}. The enzymatic site of the C3 convertase is located in the C2a molecule. After reassociation with the major split product C3b (C4b2a3b), C2a acquires enzymatic activity against C5. C4b2a is unstable and undergoes decay unless there is a sufficient quantity of C3 in the vicinity of the cell-bound complex to mediate the next site in the sequence. During this decay the C2a fragment is released in a functionally inactive form, iC2a. The remaining C4b site is now able to take up new native C2 and a new C4b2a enzyme can be generated.

The Enzymes of the Alternative Pathway

Factors B, P and D of the alternative pathway, in conjunction with C3b, the major split product of C3, generate the alternative pathway C3 convertase [12]. A variety of activators of this pathway have been described such as bacterial (LPS), yeast (zymosan), or plant (inulin) polysaccharides, polyanionic substrates like dextran sulfate, a cobra venom factor (CVF) and the Fab portions of immunoglobulins, e.g. IgA or IgE, etc.

The formation of C3bBb, the alternative pathway C3 convertase, is triggered by the spontaneous hydrolysis of the internal thioester bond in C3 which produces a molecule of C3(H_2O) possessing all the functional properties of C3b. C3(H_2O) in the presence of Mg^{++} ions associates with B to form C3(H_2O)B. Factor D then cleaves C3(H_2O)B into the convertase of C3(H_2O)Bb. The alternative pathway is in a state of continuous low grade turnover. Without the control proteins H and I, the alternative pathway is rapidly exhausted.

The control proteins serve to convert C3b and C3(H_2O) to the inactivated form of C3b designated C3bi. In the fluid phase the control protein H has a higher affinity for C3b than the affinity of B for C3b and will thus compete successfully with B for C3b. C3b in association with H or CR1, the membrane receptor for C3b, is rapidly converted by the action of I into C3b. However, on alternative pathway activating surfaces such as a viral or bacterial membrane, the affinity of H for C3b is greatly reduced allowing the positive feedback loop which characterizes the alternative pathway to take effect. Thus, on an activating surface, due to the changed relative affinities of the control protein H and factor B for C3b, alternative pathway C3 convertase is rapidly produced. Natural modulation of alternative pathway C3 convertase occurs due to the spontaneous decay of C3Bb as Bbi is released. This dissociation is retarded by the action of P which forms the stable complex C3bBbP. D exists in serum in activated form and is not itself incorporated into the convertase.

The Terminal Complement Sequence

Cleavage of C3 into C3a and C3b by the C3 convertase of the classical as well as of the alternative complement pathway and the association of C3b with these convertases leads to the formation of the C5 convertase which cleaves C5 in two fragments C5a and C5b. After cleavage of C5, the reaction C5b with C6, C7, C8, and C9 proceeds by protein-protein interactions without further proteolysis. The formed C5b-C9 complex, also called membrane attack complex (MAC), is inserted into membranes, to create transmembrane channels leading to osmotic lysis of the cell.

Regulatory Factors of the Complement System

A number of regulatory systems prevent unrestricted activation of the reaction sequence or bring about the rapid destruction of the activated factors at several stages of the complement system [6, 8, 13]. Control is exerted by either specific control proteins or the short half-life of activated components and of some of the complexes. Examples of the latter are the spontaneous dissociation of the classical pathway (C4b2a) and alternative pathway (C3b,Bb) convertases, the rapid inactivation of the labile cell binding site in C3b and C4b by water and the short life of the C6 binding site in C5b. The main regulatory functions, however, are taken over by specific control proteins, namely $C\bar{1}$-esterase inhibitor ($C\bar{1}$-INH), C4b-binding (C4bp), decay-accelerating factor (DAF), factor H, factor I, C3b-receptor (CR1), S-protein and C8-binding protein (C8bp).

In the classical pathway, C$\bar{1}$-INH controls activated C$\bar{1}$ and C1 activation [14]. C$\bar{1}$-INH is a glycoprotein of 105,000 daltons that blocks C$\bar{1}$ activity by covalent binding to the catalytic site on the light chains of C1$\bar{r}$ and C1$\bar{s}$ and by subsequent disassembly of the C$\bar{1}$ complex into two C1$\bar{r}$C1$\bar{s}$(C$\bar{1}$-INH)$_2$ complexes and C1q, which remains bound to the activator. In addition, C$\bar{1}$-INH has been reported to inhibit autoactivation of plasma C1. The classical C3 convertase (C4b2a) is controlled by DAF, a 70-kd molecule present within the membranes of many different cell types and by C4bp, a 570-kd protein. Both regulatory proteins inactivate the C4b2a complex by disassociating C2a from the convertase. In addition, DAF accelerates the decay of the alternative pathway C3 convertase and C4bp works as a cofactor for the inactivation of C4b by factor I. Factor I, also known as C3b/C4b inactivator, is a 80-kd serine esterase that cleaves C4b bound to surfaces or to C4bp in a two-step reaction to the hemolytically inactive C4c and C4b.

The terminal component lytic efficiency is restricted by S-protein and by C8bp. In the fluid phase, S-protein, an 88-kd glycoprotein, is capable of stable interactions with the membrane binding site of the C5b67 complex, thus inhibiting the attachment of the complex to the membrane. C8bp is a species-specific membrane protein that inhibits the efficiency of hemolytic lesions by both C8 and C9.

The Biological Role of the Complement Activation

The complement system is one of the most important humoral systems mediating many reactivities that contribute to inflammation and host defense even in the preimmune phase where specific antibodies and lymphocytes are not available. Therefore, it is not surprising that the complement cascades have a variety of initiators in addition to antigen-antibody reactions. Activation of the complement cascade leads to the fragmentation of C3, C4, and C5 into low molecular weight hormone-like peptides — C3a, C4a, and C5a (table 1) [15]. These molecules induce smooth muscle contraction and enhance vascular permeability. They bind to specific receptors and induce the release of vasoactive amines such as histamine from mast cells and basophils and lysosomal enzyme release from granulocytes. C5a functions also as chemoattractant inducing the migration of leukocytes into an area of complement activation.

The activities of C3a and C5a are abolished by anaphylatoxin inactivator (AI) which removes the C-terminal arginine from both molecules. The

Table 1. Biological functions of activated complement components

C2a	Kinin-like activity. Increase of vascular permeability and contraction of smooth muscles
C3a, C4a, C5a	Anaphylatoxic peptides bind to receptors on granulocytes, macrophages, mast cells and thrombocytes Release of vasoactive amines Enhanced vascular permeability Contraction of smooth muscle Induced release of lysosomal enzymes
C5a	Chemotaxis: induction of migration of leukocytes into an area of C activation Granulocyte aggregation Activation of intracellular processes such as release of oxygen metabolites and SRS-A
C3b, C4b	Immune adherence and opsonization. Bridging between a complex or target cell bearing C3b or C4b and the responding cell having a receptor for C3b or C4b: phagocytic cells (macrophages, monocytes, polymorphonuclear leukocytes), B lymphocytes, primate erythrocytes, platelets
C5b-C9	Membrane damage: lysis of cells, e.g. red cells and gram-negative bacteria

high molecular weight fragments of C3 and C4, C3b and C4b, bound to membranes of cells or bacteria, are recognized by various cells having receptors for C3b or C4b such as phagocytic cells (macrophages, monocytes, and polymorphonuclear leukocytes), B lymphocytes, neutrophils, erythrocytes, and platelets [16]. Therefore, C3b and C4b molecules serve as a bridge between a complex or target cell bearing C3b or C4b and the corresponding cell having a receptor for C3b or C4b. The consequence of C3b- or C4b-mediated bridging depends on the responding cell type: binding to phagocytic cells triggers phagocytosis. Adherence to nonprimate platelets induces specific release of vasoactive amines and nucleotides from the platelet.

Since B lymphocytes have receptors for C3b and C4b, it is postulated that bridging brings antigens in direct contact with antibody-forming cells and that, therefore, bound complement components may play a role in the induction of an immune response. Induced by activation of both complement pathways, the formation of the multimolecular C5b-C9 complex on a cell membrane results in an impairment of osmotic regulation which may cause cytolysis. The C5b-C9 complex is incorporated into the lipid layer of

Table 2. Diseases associated with complement deficiencies

C1q	Vasculitis, nephritis, systemic lupus erythematosus or similar syndrome, hypogammaglobulinemia, and chronic bacterial infections
C1r	Recurrent infections, renal disease, systemic lupus erythematosus or similar syndrome, rheumatoid disease
C1s	Systemic lupus erythematosus, Raynaud's phenomenon
C4	Systemic lupus erythematosus
C2	Systemic lupus erythematosus or similar syndrome, glomerulonephritis, vasculitis, arthralgia, susceptibility to infections
C3	Recurrent infections with pyogenic bacteria
C5	Systemic lupus erythematosus, recurrent infections (gram-negative bacteria), recurrent gonococcal infections, defect in chemotaxis
C6, C7	Recurrent meningococcal and gonococcal infections
or C8	Raynaud's syndrome, systemic lupus erythematosus, glomerulonephritis
$C\bar{1}$-INA	Hereditary angioedema, systemic lupus erythematosus-like disease
H (C3-INA)	Recurrent infections with pyogenic bacteria

the membrane leading to the formation of a transmembrane channel allowing passage of small molecules and initiating osmotic lysis of the cell. Complement-mediated lysis has been shown for many kinds of cells: erythrocytes, platelets, bacteria, viruses processing a lipoprotein envelope, and lymphocytes. The biological importance of the complement system for the maintenance of a functional host defense is impressively illustrated by the markedly increased susceptibility to infection and the predisposition to diseases observed in some congenital or acquired deficiencies of complement components or complement regulatory proteins (table 2).

Biosynthesis of Complement Components

Blood monocyte and tissue (peritoneum, lung, spleen, bone marrow) macrophages from human, guinea pig and mouse species have been cultured in vitro and all the components of the classical and alternative pathways and also the alternative pathway control proteins H and I have been shown to be synthesized. C5 is also produced but the terminal components C6-C9 are not synthesized by these cells. The presence of complement components in the macrophage supernatants was ascertained both by biochemical characterization and also functional testing.

Classical Pathway Components

C1q, C1r and C1s

C1q has a molecular weight of 460 kd and is composed of 18 polypeptide chains. There are three types of chain designated A, B and C, having molecular weights of 27.6, 25.2 and 23.8 kd, respectively. The chains are disulfide linked into 6 AB and 3 CC dimers. Viewed by electron microscopy, C1q has a structure reminiscent of a bunch of tulips. Six Fc-binding globular heads containing one of each chain type are linked by collagen-like stems to a central column. The collagen-like regions of the molecule are approximately 80 amino acids long having the typical sequence Gly-X-Y, where Y is frequently hydroxyproline or hydroxylysine. The strong similarity of C1q stem structure to collagen has allowed the use of inhibitors' collagen biosynthesis such as 3,4-dehydro-*DL*-proline and 2,2'-dipyridyl to investigate the posttranslational modification of C1q. Müller et al. [17] demonstrated the reversible inhibition of release of hemolytically active C1q from guinea pig macrophages using 2,2'-dipyridyl, an iron chelator and blocker of prolyl and lysyl hydrolases.

C1q biosynthesis has also been demonstrated in human and mouse peritoneal macrophages [18, 19]. Guinea pig peritoneal macrophages have been shown to possess a membrane form of C1q whose B chain is 1 kd less than its fluid-phase counterpart. This membrane-associated C1q is derived from already secreted C1q [20]. The membrane-associated C1q may function as a macrophage Fc-receptor [19]. Several low molecular weight (LMW) forms of C1q have been reported whose functions remain unclear. Human monocytes secrete a LMW C1q which does not associate with the $C1r_2s_2$ complex and is related to the disease systemic lupus erythematosus [21, 22]. A LMW C1q with quite different properties synthesized by guinea pig macrophages has been described by Martin and Loos [23]. This C1q retains affinity for $C1r_2s_2$ and is secreted as a large minority of total C1q by healthy outbred guinea pig peritoneal macrophages.

The C1r and C1s subcomponents of C1 are single chain serine proteinases of approximate molecular weight 85 kd. Both of these proteins are known to be synthesized by guinea pig peritoneal macrophages [17, 18] and by human blood monocytes [24]. Each of the C1 subcomponents is synthesized independently. Reid and Solomon [25] discovered that fibroblasts produce a greater quantity of the C1r and C1s subcomponents than C1q. The converse of this result was found for macrophages by Loos [19] who showed that between 15 and 30 times more functional C1q than C1 is released by guinea pig peritoneal macrophages. The C1s activity in these cultures was also greater than the C1 activity, whereas C1r activity directly corresponded to C1 function indicating that C1r is the limiting component in guinea pig peritoneal macrophage biosynthesis.

C4

The genes for components C4 and C2 of the classical pathway and factor B of the alternative pathway are encoded within the major histocompatibility system. C4 is a 200-kd polymorphic glycoprotein consisting of three disulfide-linked polypeptides. These chains are synthesized as a single chain precursor which is then cleaved during processing. Guinea pig and mouse macrophages and human blood monocytes have been shown to synthesize C4 [24, 26–28]. Although the rate of total protein synthesis by activated or elicited mouse peritoneal macrophages is greater than that of resident peritoneal macrophages, the rate of C4 biosynthesis by the activated cells is lower [29]. In contrast, factor B biosynthesis is increased in the activated macrophages. Moreover, in vitro culture of both the resident and the elicited peritoneal cells results in a decrease in C4 production.

The use of peritoneal macrophages from guinea pigs gives quite a different picture of C4 biosynthesis. The rate of C4 production in elicited guinea pig macrophages was found to be 3–4 times greater than in resident cells [30]. Thus, for C4 production by macrophages there are clear species-specific differences. Auerbach et al. [28] investigated feedback inhibition of C4 biosynthesis by guinea pig macrophages using a hemolytic plaque assay in conjunction with biosynthetic labeling and a cDNA probe for C4 mRNA. The presence of fluid-phase C4 in these macrophage cultures gave rise to a time- and dose-dependent decrease in the proportion of C4-producing cells and the reduction of the quantity of C4 mRNA. Such a feedback inhibition mechanism would allow the macrophage to respond to a depletion in C4 levels in its microenvironment with increased production until an adequate C4 level is reached. In this regard another species difference is apparent since such feedback inhibition of C4 production does not occur in mouse macrophages.

Both mouse and guinea pig macrophages vary in C4 production according to the tissues from which they are isolated [30]. This phenomenon was investigated more closely by Alpert et al. [27] using a hemolytic plaque assay.

It was found that whereas C4 was synthesized by only 10% of adherent bone marrow cells, 45% of spleen, peritoneum and lung macrophages synthesize C4. The evidence concerning C4 synthesis by human monocytes and macrophages is more controversial. Whaley [24] reported C4 synthesis by blood monocytes detected by Ouchterlony diffusion of culture supernatant. However, Cole et al. [31, 32] found no detectable C4 from blood monocytes, breast milk macrophages and bronchoalveolar macrophages as determined by hemolytic assays and immunoprecipitation of biosynthetically labeled proteins from culture supernatants and cell lysates followed by SDS-PAGE. Furthermore, initial investigations of C4 mRNA levels in human monocytes and macrophages suggest that no C4 mRNA is detectable in cells where factor B mRNA is present. Conceivably, these data may indicate that C4 production by monocytes and macrophages is not a constitutive property of these cells but rather that C4 biosynthesis is determined by the presence of local immunoregulators.

C2

C2 and factor B bear a resemblance in structure and control of biosynthesis. C2 has a molecular weight of 100 kd while factor B weighs 95 kd. On activation, both glycoproteins are cleaved into fragments of 60–70 kd (C2a and Bb) and 30–40 kd (C2b and Ba). The morphologic and histochemical maturation which human blood monocytes undergo during 16 weeks of in vitro culture were shown to correlate with changes in the rates of C2 and factor B synthesis [33, 34]. A comparison of freshly isolated monocytes with long-term monocytes, breast milk and alveolar macrophages suggested that C2 and factor B production are characteristic of differentiated macrophages. The ratio of C2 to factor B varied according to the source of the cells. After 3 days' culture of monocytes this ratio was 1:1. Breast milk and alveolar macrophages require no in vitro culture to induce production of these proteins. The C2:factor B ratio in breast milk and alveolar macrophages was 3.5:1 and 9:1, respectively. Unlike C4 production, there is no evidence for feedback inhibition of C2 synthesis.

C3

C3 has the highest serum concentration of all complement components. Synthesized as a single chain precursor, C3 is cleaved into an α-chain of 115 kd and a β-chain of 75 kd linked by disulfide bonds. Bentley et al. [35] showed that guinea pig peritoneal macrophages produce C3, and Zimmer et al. [36] demonstrated that starch gel stimulation of peritoneum results in a 3-

fold increase in C3 synthesis by macrophages. Human monocytes and macrophages from human lung and mouse peritoneum are also known to produce C3 [24, 32, 37, 38].

The Alternative Pathway

Factors B, D and P of the alternative pathway have been shown to be secreted by human monocytes [24, 39] and guinea pig and human macrophages [40–42].

Factor B

Mouse peritoneal macrophages synthesize factor B [35]. Treatment of guinea pig macrophages with tunicamycin, an antibiotic which inhibits glycosylation of proteins, blocked secretion of factor B [43]. Factor B synthesis appears to correlate positively with macrophage maturation. Whaley [24] observed that, during the first 25 h of culture, human monocyte factor B synthesis was minimal and that during the following 3 days there was a rapid increase in the biosynthetic rate. Cole et al. [31, 32] found that unlike with blood monocytes, there is no lag phase in the production of factor B from freshly isolated human bronchoalveolar and breast milk macrophages. Factors known to induce increased factor B synthesis by macrophages include LPS, thioglycolate stimulation in vivo, anti-C3 Fab', interferon-γ and interleukin-1.

Factor D

Factor D biosynthesis by macrophages varies from species to species. Guinea pig macrophages produce sufficient D to be detectable by functional assay [44], whereas no D synthesis by mouse macrophages is apparent [35]. In one report, D synthesis by human monocytes was induced by long-term in vitro culture and this appeared to be a function associated with monocyte/macrophage differentiation [24]. In contrast, other reports [33] suggest that there is no delay in D synthesis by freshly isolated human monocytes.

Properdin

Properdin (P) synthesis by guinea pig macrophages has been demonstrated by immune precipitation of radiolabeled protein [44], and functional tests [41, 42]. Human monocytes and macrophages have been shown to synthesize functionally active P [24, 39].

C5

C5 is the only component of the terminal sequence known to be synthesized by macrophages. In its native form, C5 consists of two chains having a total molecular weight of 180 kd. However, biosynthetic labeling of mouse peritoneal macrophages revealed that C5 is synthesized as a single chain precursor which is intracellularly cleaved into the native C5 form [45]. There is strong evidence that posttranslational modification of C5 is a critical step in the control of C5 function: Ooi et al. [46] found that mouse macrophages elicited by thioglycolate stimulation of the peritoneum produced twice as much C5 antigen as resident cells. However, the thioglycolate-elicited cells produced approximately only 20% of the C5 functional activity produced by the resident cells. C5 biosynthesis also differs from various other complement proteins in its response to histamine. In contrast to the elevated production of C2, C3, C4, B and H induced by histamine treatment of mouse peritoneal macrophages, C5 synthesis is depressed as determined both functionally and antigenically [47].

Factors H, I and C$\bar{1}$-INH

Whaley [24] demonstrated the synthesis of functionally detectable factor I and antigenically detectable factor H by human peripheral blood monocytes. C$\bar{1}$-INH biosynthesis by mononuclear phagocytes was shown by Bensa et al. [48]. Yeung Laiwah et al. [49] showed that C$\bar{1}$-INH biosynthesis is controlled independently of other complement components. After 7 days in vitro culture of human monocytes C2 biosynthesis ceased, whereas C$\bar{1}$-INH production continued. Reversible inhibition by cycloheximide confirmed C$\bar{1}$-INH biosynthesis using human monocytes. Lappin et al. [50] found that C$\bar{1}$-INH is clearly an inducible protein since biosynthesis did not occur until 3 days' in vitro culture. C$\bar{1}$-INH production along with C2 and factor B is stimulated by recombinant *Escherichia coli* derived interferon-γ although C3 production is diminished [51].

Regulation of Biosynthesis

Clearly there are wide variations among tissues in the percentage of macrophages which synthesize a given complement component. In addition, within those cells synthesizing a component there is also a wide range of biosynthetic rates. To some extent these variations may compensate for one another to yield a more even overall level in the production of complement components by macrophages in different tissues. For example, in the guinea pig, 45% of splenic and peritoneal macrophages synthesize C2, whereas C2 is produced by only 2.5% of lung macrophages. However, on an individual cell basis the C2-producing lung macrophages have a biosynthetic rate of C2 5–10 times greater than that of the peritoneal macrophages [27, 52]. Thus, the net C2 production by macrophages in the peritoneum and the lung is to some extent balanced. It is as yet unknown whether these variations can be accounted for on the basis of selective migration of complement synthesizing macrophages into tissues or whether the tissue microenvironment itself controls the macrophage's differentiation after its arrival.

The presence of soluble cytokines plays a role in control of C2 secretion by human monocytes. Littmann and Ruddy [53] showed that the initial lag in C2 biosynthesis which occurs on in vitro culture of human monocytes is abolished by addition of a secreted factor from an antigen-stimulated lymphocyte culture. Modulation of C3 biosynthesis by human monocytes by lipid A has been demonstrated by Strunk et al. [54]. Lipid A treatment of these cells induced increased C3 production by up to 30-fold, while C2 and factor B biosynthesis remained virtually unaffected. This effect was associated with an increase in C3 mRNA. Curiously, C2 and factor B mRNA levels also increased but these effects did not correspond to changes in the C2 and B protein levels. These data suggest that a second level of control is exerted on the specific translation rates of the individual mRNAs and that increased mRNA alone is not enough to enhance protein production.

Concluding Remarks

Viewing the variability of complement component production by macrophages among species and also among tissues, one receives the impression that general statements covering this topic are difficult to make. Thus, the cells which produce a component may show species and tissue variation so long as a functional level of the component is achieved. The mechanisms which operate at the levels of mRNA synthesis, mRNA processing and translation, and finally posttranslational modification to influence the functional activity of the components are currently being elucidated. The recent successes in cloning complement genes allow these areas to be investigated. The analysis of complement biosynthesis has yielded information with general implications for the regulation and modulation of gene expression.

This is of particular relevance for macrophages which are migratory, invasive cells subject to the influence of many different microenvironments. One such example is the independent regulation of C2 and factor B genes within the major histocompatibility complex which are separated by only 500 base pairs.

References

1 Ruddy S, Colten HR: Rheumatoid arthritis. Biosynthesis of complement proteins by synovial tissues. N Engl J Med 1974;290:1284–1288.

2 Cooper NR: The classical complement pathway: Activation and regulation of the first complement component. Adv Immunol 1985;37:151–216

3 Fothergill JE, Anderson WHK: A molecular approach to the complement system. Curr Top Cell Regul 1980;13:259–311.

4 Müller-Eberhard HJ: Complement. Annu Rev Biochem 1975;44:697–724.

5 Müller-Eberhard HJ, Schreiber RD: Molecular biology and chemistry of the alternative pathway of complement. Adv Immunol 1980;29:1–53.

6 Müller-Eberhard HJ: Chemistry and function of the complement system; in Dixon FJ, Fisher DW (eds): The Biology of Immunologic Disease. Sunderland, Sinauer Assoc, 1983.

7 Porter RR, Reid KBM: Activation of the complement system by antibody-antigen complexes: the classical pathway. Adv Protein Chem 1970;33:1–71.

8 Ross GD: Immunobiology of the Complement System. New York, Academic Press, 1986.

9 Loos M: The classical complement pathway: Mechanism of activation of the first component by antigen-antibody complexes. Prog Allergy. Basel, Karger, 1982, vol 30, pp 135–192.

10 Loos M: Antibody-independent activation of C1, the first component of complement. Ann Immunol (Paris) 1982;133C:165–179.

11 Loos M, Heinz HP: Generation of the classical pathway C3 convertase (EAC4b2a) by proteolytic enzymes. Acta Pathol Microbiol Immunol Scand Sect C 1984;92 (suppl 284): 67–74.

12 Pangburn MK, Müller-Eberhard HJ: The alternative pathway of complement. Springer Semin Immunopathol 1984;7:163–192.

13 Fries LF III, Frank MM: Molecular mechanisms of complement action; in Stamatoyannopoulos G, Nienhuis AW, Leder P, et al (eds): The Molecular Basis of Blood Diseases. Philadelphia, Saunders, 1987.

14 Schapira M, de Agostini A, Schifferli JA, et al: Biochemistry and pathophysiology of human C$\bar{1}$ inhibitor: Current Issues. Complement 1985;2:111–126.

15 Hugli TE: Structure and function of the anaphylatoxins. Springer Semin Immunopathol 1984;7:193–219.

16 Schreiber RD: The chemistry and biology of complement receptors. Springer Semin Immunopathol 1984;7:221–249.

17 Müller W, Hanauske-Abel H, Loos M: Reversible inhibition of C1q release from guinea pig macrophages by 2,2'-dipyridyl. FEBS Lett 1978;90:218–222.

18 Müller W, Hanauske-Abel H, Loos M: Synthesis of the first component of complement by human and guinea pig macrophages: evidence for an independent production of the C1 subunits. J Immunol 1978;121:1578–1584.
19 Loos M: Biosynthesis of the collagen-like C1q molecule and its receptor functions for Fc and polyanionic molecules on macrophages. Curr Top Microbiol Immunol 1983;102:1–56.
20 Martin H, Heinz H-P, Reske K, et al: Macrophage C1q: Characterization of a membrane form of C1q and of multimers of C1q subunits. J Immunol 1987; 138:3863–3867.
21 Hoekzema R, Brouwer MC, Hack CE: Structural studies on low molecular weight C1q (abstract). Complement 1985;2:35.
22 Hoekzema R, Hannema AJ, Swaak TJG, et al: Low molecular weight C1q in systemic lupus erythematosus. J Immunol 1985;135:265–271.
23 Martin H, Loos M: Guinea-pig macrophages synthesize a low molecular weight form of C1q with affinity for the $C1r_2$ $C1s_2$ complex but which does not bind Fc in immunoglobulin aggregates. Mol Immunol, in press.
24 Whaley K: Biosynthesis of the complement components and the regulatory proteins of the alternative complement pathway by human peripheral blood monocytes. J Exp Med 1980;151:501–516.
25 Reid KBM, Solomon E: Biosynthesis of the first component of complement by human fibroblasts. Biochem J 1977;167:647–660.
26 Newell SL, Shreffler DC, Atkinson JP: Biosynthesis of C3 by mouse peritoneal macrophages. I. Characterization of an in vitro culture system and comparison of C4 synthesis of 'low' vs. 'high' C4 strains. J Immun 1982;129:653–659.
27 Alpert SE, Auerbach HS, Cole FS, et al: Macrophage maturation: differences in complement secretion by marrow, monocyte, and tissue macrophages detected with an improved hemolytic plaque assay. J Immunol 1983;130:102–107.
28 Auerbach HS, Lalande ME, Latt S, et al: Isolation of guinea pig macrophages bearing surface C4 by fluorescence activated cell sorting: correlation between surface C4 antigen and C4 protein secretion. J Immunol 1983;131:2420–2426.
29 Newell SL, Atkinson JP: Biosynthesis of C4 by mouse peritoneal macrophages. II. Comparison of C4 synthesis by resident and elicited cell populations. J Immunol 1983;130:834–838.
30 Cole FS, Matthews WJ, Marino JT, et al: Control of complement synthesis and secretion in bronchoalveolar and peritoneal macrophages. J Immunol 1980; 125:1120–1124.
31 Cole FS, Schneeberger EE, Lichtenberg NA, et al: Complement biosynthesis in human breast milk macrophages and blood monocytes. Immunology 1982;46:429–441.
32 Cole FS, Matthews WJ, Rossing TH, et al: Complement biosynthesis by human bronchoalveolar macrophages. Clin Immunol Immunopathol 1983;27:153–159.
33 Beatty DW, Davis AE, Cole FS, et al: Biosynthesis of complement by human monocytes. Clin Immunol Immunopathol 1981;18:334–343.
34 Einstein LP, Schneeberger EE, Colten HR: Synthesis of the second component of complement by long-term primary cultures of human monocytes. J Exp Med 1976;143:114–126.
35 Bentley CD, Bitter-Suermann D, Hadding U, et al: In vitro synthesis of factor B of the alternative pathway of complement by mouse peritoneal macrophages. Eur J Immunol 1976;6:393–398.

36 Zimmer B, Hartung HP, Scharfenberger G, et al: Quantitative studies of the secretion of complement components C3 by resident, elicited, and activated macrophages. Comparison with C2, C4 and lysosomal enzyme release. Eur J Immunol 1982;12:426–430.
37 Strunk RC, Kunke KS, Giclas PC: Human peripheral blood monocyte-derived macrophages produce haemolytically active C3 in vitro. Immunology 1983;49:169–174.
38 Fey G, Domdey H, Wiebauer K, et al: Structure and expression of the C3 gene. Springer Semin Immunopathol 1983;6:119–147.
39 De Ceulaer C, Papazoglon S, Whaley K: Increased biosynthesis of complement components by cultured monocytes, synovial fluid macrophages and synovial membrane cells from patients with rheumatoid arthritis. Immunology 1980;41: 37–43.
40 Bentley CD, Zimmer B, Hadding U: The macrophage as a source of complement components; in Pick E (ed): Lymphokines. New York, Academic Press, 1981, vol 4, pp 197–230.
41 Brade V, Bentley CD: Synthesis and release of complement components by macrophages; in van Furth R (ed): Mononuclear Phagocytes. Functional Aspects. The Hague, Nijhoff, 1980, pp 1385–1417.
42 Brade V, Kreuzpaintner G: Functional active complement components secreted by guinea pig peritoneal macrophages. Immunobiology 1982;161:315–321.
43 Matthews WJ, Goldberger G, Marino JT, et al: Complement proteins C2, C4, and factor B: effect of glycosylation on their secretion and catabolism. Biochem J 1982;204:839.
44 Bentley CD, Fries W, Brade V: Synthesis of factors D, B and P of the alternative pathway of complement activation as well as C3 by guinea pig peritoneal macrophages in vitro. Immunology 1978;35:971–980.
45 Ooi YM, Colten HR: Biosynthesis and post-synthetic modification of a precursor (pro-C5) of the fifth component of mouse complement (C5). J Immunol 1979;123:2494–2498.
46 Ooi YM, Harris DE, Edelson PJ, et al: Post-translational control of complement (C5) production by resident and stimulated mouse macrophages. J Immunol 1980;124:2077–2081.
47 Ooi YM: Histamine suppressed in vitro synthesis of precursor (pro-C5) of the fifth complement component (C5) by mouse peritoneal macrophages. J Immunol 1982;129:200–205.
48 Bensa JC, Reboul A, Colomb MG: In vitro biosynthesis of C1 subcomponents and $C\bar{1}$-INH by human monocytes (abstract). Immunobiology 1983;164:210.
49 Yeung-Laiwah AC, Jones L, Hamilton AO, et al: Complement subcomponent $C\bar{1}$-inhibitor synthesis by human monocytes. Biochem J 1985;226:199–205.
50 Lappin D, Hamilton AD, Morrison L, et al: Synthesis of complement components (C3, C2, B and $C\bar{1}$ inhibitor) and lysozyme by human monocytes and macrophages. J Clin Lab Immunol 1986;20:101–105.
51 Hamilton AO, Jones L, Morrison L, et al: Modulation of monocyte synthesis by interferons. Biochem J 1987;242:809–815.
52 Cole FS, Auerbach HS, Goldberger G, et al: Tissue-specific pretranslational regulation of complement production in human mononuclear phagocytes. J Immunol 1985;134:2610–2616.

53 Littmann BH, Ruddy S: Production of the second component of complement by human monocytes: stimulation by antigen activated lymphocytes or lymphokines. J Exp Med 1979;145:1344–1352.
54 Strunk RC, Whitehead AS, Cole FS: Pretranslational regulation of the synthesis of the third component of complement in human mononuclear phagocytes by the lipid A portion of lipopolysaccharide. J Clin Invest 1985;76:985–990.

Dr. Michael Loos, Institut für Medizinische Mikrobiologie, Johannes-Gutenberg-Universität, Hochhaus am Augustusplatz, D-6500 Mainz (FRG)

Sorg C (ed): Macrophage-Derived Cell Regulatory Factors.
Cytokines. Basel, Karger, 1989, vol 1, pp 173–192

Monocytes and Granulomatous Inflammatory Responses

Sharon M. Wahl

Cellular Immunology Section, National Institute of Dental Research,
National Institutes of Health, Bethesda, Md., USA

Introduction

Granulomatous inflammatory responses reflect the host's attempt at defense against persistent foreign substances that have resisted destruction by the acute inflammatory response. The primary requisite for the generation of a granulomatous response is the occurrence of sustained tissue irritation by poorly degradable foreign substances. Such a substance stimulates continued phagocyte mobilization and activation leading to tissue destruction and, eventually, tissue repair. The purpose of this organized host response is to sequester and degrade the irritant. If the irritant is infectious, failure to generate an adequate granulomatous response can result in rapid dissemination of the microorganisms. Although meant to be protective, the granulomatous response may, in fact, become destructive [1–3]. The outcome of granulomatous inflammation can vary widely from nearly complete resolution with minimal scarring to excessive fibrosis causing irreversible damage to the tissues.

Granulomatous inflammatory responses can be characterized as nonimmunologic responses to relatively inert, nondigestible particles (silica, carageenan, sutures, etc.) or as immunologically induced hypersensitivity responses to infectious (table 1) or other poorly degradable antigenic substances [1]. The antigens inducing several types of granulomatous disorders are unknown as in sarcoidosis and Wegener's granulomatosis in the lungs, Crohn's disease in the intestine, and primary biliary cirrhosis. Whereas nonimmunologic granulomas are composed primarily of macrophages and fibroblasts, immune-based granulomas, evoked by slowly degradable, particulate or slowly released soluble antigens are more complex with T and B

Table 1. Granulomatous diseases caused by infectious organisms

Bacteria	*Fungus*
Tuberculosis	Histoplasmosis
Leprosy	Blastomycosis
Brucellosis	Coccidiomycosis
Listerosis	Hypersensitivity pneumonitis
Syphilis	
Q fever	*Helminths*
Cat scratch fever	Schistosomiasis
Lymphogranuloma venereum	Trichinosis
Virus	
Cytomegalovirus	

lymphocyte participation (fig. 1). However, macrophages are the basic constituents of granulomatous lesions whether of foreign body or immunologic etiology.

Until the 1970s, it was not appreciated that the primary cell constituting various forms of granulomas was of the monocyte lineage [1–3]. However, since that time, the role of the monocyte-macrophage in the evolution of granuloma development has been extensively studied. It is now recognized that ingestion of the inciting agent by the mononuclear phagocytes and persistence of the poorly degradable agent causing continued macrophage activation are responsible for the evolution of a chronic granulomatous inflammatory response. Although much has been learned concerning granuloma development since the original descriptions of tuberculosis, considered the prototypic granulomatous disease [1], the recruitment mechanisms, network of cell-cell interactions, molecular signals, role of cytokines, and the pathogenesis of associated destructive or fibrotic disease are not completely defined.

The development of a granuloma begins as any other inflammatory response when the host encounters an antigenic stimulus or foreign invader. An acute inflammatory response is initiated which is characterized by increased vasodilation and vascular permeability, edema, the release of a variety of inflammatory mediators, and the infiltration of polymorphonuclear leukocytes (fig. 1). As these neutrophils attempt to engulf the foreign material, they in turn release an additional battery of inflammatory mediators associated with the infiltration of mononuclear cells (fig. 1). If the foreign material is not readily degradable and persists in the tissue, continued turnover and recruitment of mononuclear phagocytes occurs. Crucial to the

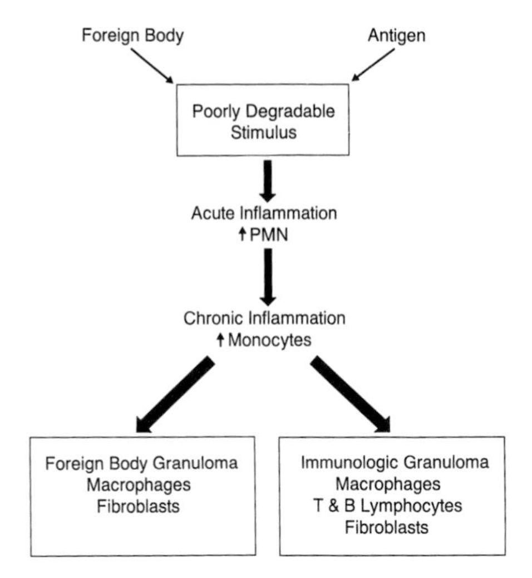

Fig. 1. Summary of nonimmunologic and immunologic granuloma development.

development of a localized, mononuclear cell-dependent granuloma is the continued activation of the cells with the secretion of products which amplify and perpetuate the lesions. Resolution of such a localized persistent inflammatory response is dependent on elimination of the antigen and/or generation of a fibrotic connective tissue barrier which serves to sequester the antigenic stimulus.

Experimental Granuloma Model

In recent studies, we have defined an experimental model of granulomatous inflammation in which we have evaluated the contribution of mononuclear cells and their products to the sequelae of events culminating in establishment of organized granulomas and hepatic fibrosis. Experimental models in which stimulus and onset of response are defined are useful for analyzing the kinetics and relationships of cell-cell interactions responsible for granuloma formation. In our model, granulomas were initiated within the livers of susceptible rat strains by a single intraperitoneal injection of group A streptococcal cell walls (SCW) [4, 5]. Whereas intact streptococci evoke hepatic granulomas following dermal or intravenous injection [6], it is the residual undegraded bacterial cell walls which persist and induce the granulomatous inflammation. The active component of the cell walls is a complex of mucopeptide and C polysaccharide [7] in which the polysaccharide protects the peptidoglycan from enzymatic degradation. This undegraded complex is ingested and retained in tissue macrophages. Lack of adequate breakdown of the polysaccharide-peptidoglycan complex is the main factor in the chronicity of the hepatic inflammation. The SCW serve as a continuous antigenic stimulus for T lymphocytes [4] provoking cell-mediated immune interactions, cytokine generation and augmented monocyte-macrophage functions. The course of the ensuing granulomatous reaction is dependent on the duration and severity of the host response to the SCW. The host response to the SCW appears to be genetically determined since certain strains of rats are susceptible to the development of chronic inflammatory lesions and others are resistant [8]. Although resistant and susceptible rats both develop an acute response to the antigen, the intensity and duration of the inflammation, evolution into a chronic granulomatous response, and speed of resolution appear to be under genetic influence.

Cell Recruitment

Hepatic granuloma formation in response to SCW is initiated once the SCW localize in the phagocytic Kupffer cells within the liver sinusoids. Shortly thereafter, leukocytes accumulate within the sinusoids and adhere to the endothelium. As the leukocytes move into the extravascular space, likely as a consequence of the release of chemotactic complement cleavage fragments [9], platelet factors [10], proteolytic peptides derived from C-reactive protein [11], and arachidonic acid metabolites [12], these cells begin to congregate in the parenchymal tissues. Within the first 1–4 days, the accumulating leukocytes are represented primarily by neutrophils, some mast cells and eosinophils (fig. 2). Mononuclear cells then begin to emigrate from the sinusoids and periportal vessels becoming predominant in the developing cellular aggregates within the liver parenchyma. The neutrophils, unsuccessful in eliminating the SCW, are likely instrumental in the subsequent influx of mononuclear cells since the neutrophils precede mononuclear cell infiltration and pharmacologic agents which inhibit the acute neutrophilic response also inhibit mononuclear leukocyte recruitment [13]. By immunoperoxidase staining with an antibody to SCW, it is apparent that the aggregates of leukocytes are associated with the localized deposition of SCW identified both extracellularly and intracellularly (fig. 3A, C). The SCW persist as the

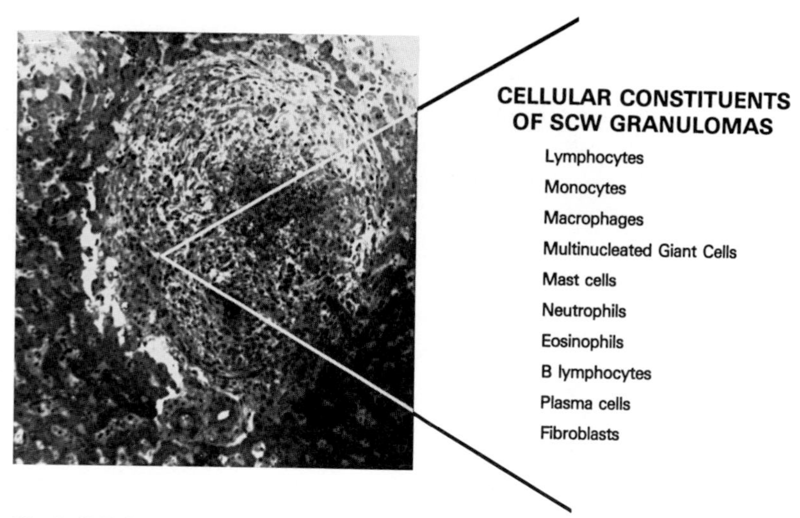

Fig. 2. Cellular constituents of SCW-induced hepatic granulomas.

focus for the organizing mononuclear cells. Since the SCW are not readily degradable, they are stored within the macrophages for long periods of time promoting chronic cell-mediated immune phenomena responsible for the granulomatous lesions [4, 5].

During the course of the next 2–3 weeks, the mononuclear cells continue to organize into compact, circumscribed granulomatous lesions (fig. 2). Identification of the cellular constituents by immunoperoxidase staining using monoclonal antibodies directed at mononuclear cell subsets revealed significant numbers of T lymphocytes primarily of the $CD4^+$ helper/inducer subset with a lesser complement of $CD8^+$ suppressor/cytotoxic cells. The influx of helper/inducer phenotype T cells correlated with the period of peak inflammation in the livers between 3 and 6 weeks. Activation of T cells is an important amplification step in SCW-induced granuloma formation. Once activated by SCW in association with Ia^+ accessory cells, the T cells proliferate and generate lymphokines which regulate the proliferative immune response and are crucial to the development of the chronic inflammatory lesions [4]. In later stage granulomas, $CD8^+$ lymphocytes increase, and these dynamic changes in lymphocyte subsets may be decisive in the evolution and involution of the lesions. B lymphocytes are also present during granuloma development; however, antibody production and their contribution to the lesions have not yet been evaluated in this model.

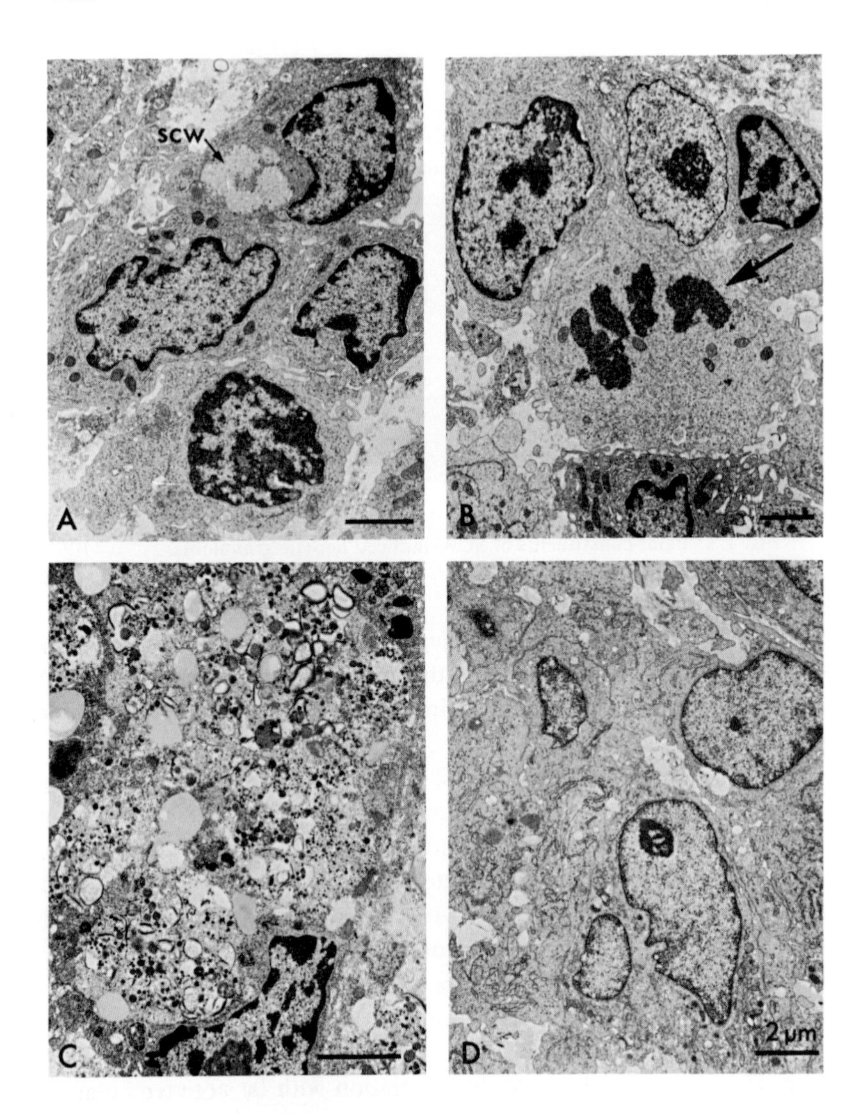

Fig. 3. Monocyte maturation in SCW-induced hepatic granulomas. *A* Immature monocytes are recruited from the circulation into the granuloma where they phagocytize SCW. *B* The monocytes may undergo a round of division (arrow) after arriving at the inflammatory site. *C* Recruited monocytes are stimulated to mature into highly phagocytic macrophages with extensive intracytoplasmic vacuoles and organelles. *D* Further maturation of the macrophages results in the development of closely intertwined epithelioid cells.

Compelling evidence for the central role of monocytes is their continued presence at all stages of the granulomas. Monocytes identified by positive Ia staining as well as other monocyte specific antigens are clearly a major constituent of the hepatic granulomas. Earlier studies characterizing cell traffic into and out of evolving granulomatous lesions using radiolabeled leukocytes demonstrated continued recruitment of circulating cells from the vasculature [14, 15] and monocytes with features consistent with a newly recruited population are seen continuously. Newly recruited monocytes (fig. 3A) may undergo a round of division after entering the lesion (fig. 3B) where they ingest SCW (fig. 3C) and become activated. Upon activation, the macrophages exhibit expanded cytoplasm, abundant Golgi, dilated endoplasmic reticulum, many mitochondria and numerous lysosomes and phagocytic vacuoles (fig. 3C). Such differentiating phagocytes characteristically have increased respiratory burst activity, protein synthesis, phagocytosis and microbicidal function. Further maturation of the mononuclear phagocytes results in the appearance of epithelioid cells (fig. 3D), closely intertwined large cells with complex, extensive cytoplasm packed with organelles. Fusion of macrophages and/or epithelioid cells results in the formation of multinucleated giant cells (fig. 4B) whose primary function is unknown, but which may retain certain monocytic functions. The inflammatory foci are in a dynamic state with continued ingress and egress of cells. Eventually, the involuting granulomas, surrounded by fibroblasts and matrix, demonstrate diminished numbers of immature monocytes with the pace of cellular turnover slowing down. Long-lived macrophages are apparent in these late lesions.

Monocytes and Their Products

Monocytes exposed to SCW in vitro synthesize and secrete a variety of inflammatory mediators which may be relevant to the in vivo inflammatory response induced by these bacterial products [16]. Concentrations of SCW greater than 0.01 μg/ml SCW stimulated macrophage Ia expression and the production of interleukin-1 (IL-1) and tumor necrosis factor (TNF). Additionally, SCW preparations stimulate macrophage oxygen metabolism and the release of toxic oxygen species including superoxide anion (O_2^-) [17]. Exposure to SCW in vitro up-regulates arachidonic acid metabolism, and the release of lipoxygenase and cyclooxygenase products may contribute to vascular and inflammatory events in the liver (table 2). Importantly, the

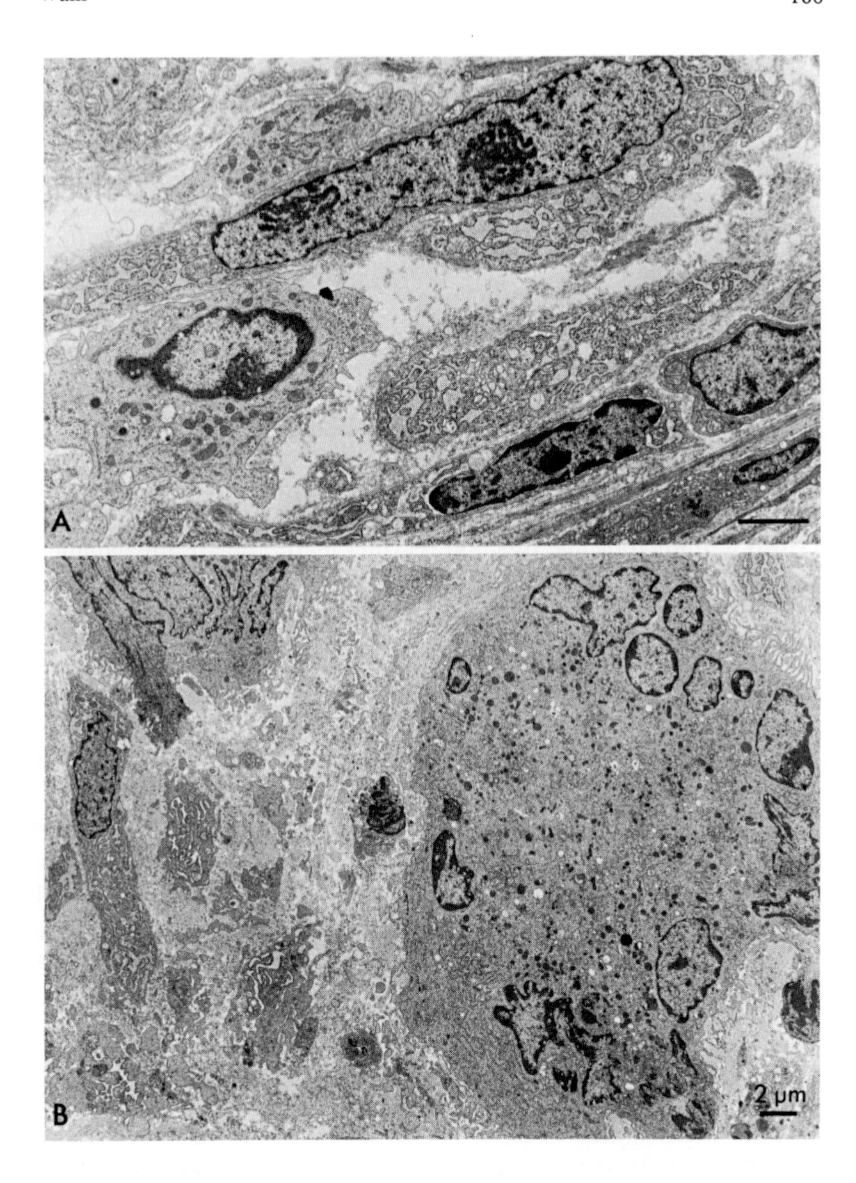

Fig. 4. Macrophage regulation of fibrosis. *A* Macrophages are found in close proximity to synthetically active fibroblasts and newly deposited matrix in the periphery of the granuloma. *B* The contribution of multinucleated giant cells located among the activated fibroblasts to fibroblast proliferation and matrix synthesis is unknown.

Table 2. SCW-induced monocyte products important in inflammation and fibrosis

Reactive oxygen intermediates (O_2^-, H_2O_2)
Arachidonic acid metabolites
Cyclooxygenase products: PGE_2, PG, 6-keto-PGF_1
Lipoxygenase products: LTC4
Interleukin-1
Tumor necrosis factor α
Fibroblast-activating factor
Transforming growth factor β
Fibroblast growth factor[1]
Platelet-derived growth factor[1]
Granulocyte-macrophage colony-stimulating factor
Collagenase

[1] Biologic activity not yet identified.

peptidoglycan-polysaccharide complexes persist within the macrophages providing chronic stimulation of these cells.

Since SCW clearly stimulated macrophages to generate inflammatory mediators in culture, subsequent experiments focused on whether these products were associated with the evolving granulomas. By 3 weeks after the initial injection of SCW, the mononuclear cells have formed discrete granulomas within the liver which are clearly delineated from the surrounding parenchymal tissue. Because of their compact, dense nature it is possible to isolate the granulomas intact from the liver tissue for the purposes of culture or isolation of the mononuclear cells by enzymatic dispersal. Culture of the isolated granulomas and/or constituent mononuclear cells documented the in situ activation of these populations by both morphologic and functional parameters. Isolated granuloma-derived macrophages were enlarged and contained increased numbers of intracellular organelles and phagolysosomes typical of an activated population. Moreover, mononuclear phagocytes obtained from the granulomas constitutively released numerous inflammatory products consistent with those produced by control monocytes exposed to SCW in vitro (table 2). Without any exogenous stimulus in vitro, these granuloma-derived macrophages released significant levels of prostaglandins (PGE_2) into the culture supernatants. Prostaglandins and other arachidonic acid metabolites are involved in a complex series of interactions with many cell types, both pro-inflammatory and anti-inflammatory.

One function of prostaglandins may be to modulate cytokine production and increasing PGE_2 synthesis was associated with decreasing IL-1 levels in maturing granulomas [4]. In this regard, IL-1 levels were highest in the early granulomas (3 weeks) at the time many newly recruited monocytes were infiltrating the granulomas and during T cell activation. Increased IL-1 production and MHC Class II (Ia) expression are important components of the role of monocytes as accessory cells in processing and promoting SCW activation of T lymphocytes. IL-1 spontaneously elaborated by granuloma cultures, is likely a key macrophage product through the course of granulomatous inflammation. As reviewed in Chapter 4 [18], IL-1 is not only an immunomodulatory protein, but also is instrumental in angiogenic and fibrotic responses as occur in these lesions (see below). However, as the granulomas mature (9–12 weeks), IL-1 levels declined. The decrease in detectable IL-1 may be the consequence of down-regulation by concomitant increasing PGE_2 levels, the reported decline in IL-1 production associated with maturation of monocyte populations [19, 20], and/or the presence of suppressor molecules. Other monokines identified in the granulomas include TNF, transforming growth factor beta (TGF-β), and granulocyte-macrophage colony-stimulating factor (GM-CSF) (table 2), all of which contribute to the network of cellular interactions necessary to maintain these chronic inflammatory responses.

Activated monocyte-macrophages from granulomas likely generate many other proteins and lipid molecules which contribute to the activation of other cell populations and the chronicity of these lesions [21]. The SCW-laden macrophages which cannot divide retain residual SCW until they die. The SCW are released, taken up by other macrophages which are activated to secrete more recruitment and activation factors, and the response becomes self-perpetuating.

Role of T Cells

Although SCW can directly activate monocyte-macrophages to secrete a plethora of enzymes, monokines and other inflammatory products, this SCW-monocyte interaction does not appear to be sufficient to sustain a granulomatous inflammatory response. In a series of studies, we have documented the absolute necessity for T cell participation in the evolution of hepatic granulomas [4, 5, 22, 23]. Although lymphocytes are key elements in the formation of immunologic granuloma as shown for schistosome granulomas [24, 25], they are frequently absent in foreign body granulomas [1]. Earlier studies favored a foreign body mechanism for SCW-induced granulomatous inflammation [1], whereas recent evidence indicates the immunologic nature of the response. In athymic, T cell-deficient animals, the injection of SCW does not provoke the formation of hepatic granulomas [22]. Even though SCW localize to this target tissue, and the host initiates an inflammatory response to the hepatic deposition of the SCW, this response is muted and full-blown granulomatous lesions never develop in the athymic rat livers. In related and confirmatory studies, animals treated with the immunosuppressive agent, ciclosporin A, which inhibits proliferation and lymphokine production by T cells, likewise were unable to sustain an inflammatory response sufficient for the development of hepatic granulomas [4]. However, in both cases, this absence of a granulomatous response could not be attributed to lack of SCW deposition in the liver nor to a failed acute inflammatory response to the SCW. Consequently, although the evidence favors a T cell-independent acute inflammatory response, the ensuing chronic mononuclear cell-mediated formation of hepatic granulomas requires SCW activation of T lymphocytes, the generation of lymphokines, and the recruitment and activation of monocytes.

In such a cell-mediated immune reaction, the lymphokine network can exert powerful inflammatory effects. This concept is supported by the demonstration that SCW antigens (peptidoglycan moiety of SCW) activate T lymphocytes in vitro to both proliferate and to secrete lymphokines, and more convincingly, by the constitutive elaboration of lymphokines by $CD4^+$ T lymphocytes isolated from SCW-induced hepatic granulomas [4, 5]. These lymphocyte populations, in the absence of exogenous stimulation, release lymphokines that can recruit and/or activate monocyte populations. In this regard, granuloma-derived lymphocytes secrete lymphocyte-derived chemotactic factor (LDCF) and granulomas produce TGF-β, both important chemotactic signals for monocytes [4, 26]. Following recruitment, T cell products, in particular gamma-interferon (γ-IFN), GM-CSF, IL-2 and TGF-β, are known to contribute to the immobilization and maturation of the monocytes within the inflammatory site. Colony-stimulating factors, polypeptides which regulate the proliferation and differentiation of mononuclear phagocytes and are constitutively produced by the granulomas [4], are found early in granuloma development and may promote mononuclear cell proliferation locally (fig. 3B). Early generation of GM-CSF in the granulomas is likely due to SCW-activated T lymphocytes, but the continued release of this cytokine, even as the T cells decline in number ($>$12 week granulomas), may be from activated macrophages [27] and infiltrating fibroblasts [28]. The availability of recombinant GM-CSF has enabled studies demonstrating that this cytokine can regulate differentiated monocyte function in addition to its anticipated proliferative effects on hematopoietic precursors. Both GM-CSF and γ-IFN induce monocyte differentiation and activation as evaluated by enhanced Ia expression [29, 30]. Lymphokine-activated macrophages display increased phagocytic and digestive properties as well as enhanced accessory cell function. Furthermore, when exposed to γ-IFN or GM-CSF, monocytes transcribe, translate and express IL-2 receptors [31], rendering them susceptible to further T cell regulation via IL-2. Once monocytes express IL-2 receptors, IL-2 will evoke reactive oxygen intermediate generation, enhanced cytotoxic activity [31] and increased production of IL-1 [32]. The complex network of facilitatory and inhibitory lymphokines likely determines the dynamics and extent of the host response, and the tight organizational structure of the lymphocytes and macrophages which develops effectively walls off the SCW from the surrounding tissues minimizing further dissemination of the residual antigen.

Angiogenesis and Fibrosis

Persistent activation of mononuclear cells and the continued release of cytokines impacts on target cells in addition to those involved in host defense against the foreign agent. Immune cell products modulate the behavior of neighboring endothelial cells in the initiation of an angiogenic response. Recruitment and proliferation of endothelial cells necessary for new capillary growth is associated with mononuclear cell accumulation. Mononuclear cell products which contribute to neovascularization processes include PDGF, fibroblast growth factor (FGF), TNF and TGF-β [33]. Although TGF-β is a potent angiogenic agent in vivo, it appears to inhibit endothelial cell proliferation in vitro. This paradox may be explained, in part, by the ability of TGF-β to recruit monocytes and to stimulate them to produce growth factors relevant for angiogenesis [26], thereby promoting these events indirectly. Recruitment and proliferation of endothelial cells results in the formation of capillary sprouts which anastomose and form capillary beds necessary for the establishment of blood flow in the developing granulation tissue [33].

Dependent upon and accompanying these events is the noticeable infiltration of fibroblasts around the periphery of the focal mononuclear cell core (fig. 4). Granulomatous lesions are often characterized by large numbers of fibroblasts which have been recruited and induced to proliferate by products of the inflammatory cells. Fibroblasts, migrating into the injury site from neighboring connective tissue, respond to chemotactic products of platelets (PDGF, TGF-β), neutrophils (LTB4), T lymphocytes (LDCF-F), and monocytes (TGF-β, fibronectin, PDGF) [34]. Fibroblasts are important to the outcome of chronic inflammation since matrix synthesis may terminate the inflammation, but unfortunately, may also cause irreversible tissue pathology.

Fibrosis, which is a repair process initiated by fibroblasts, normally follows tissue injury. The process includes recruitment of fibroblasts into the area of injury, local proliferation of the fibroblasts, collagen and matrix synthesis, and tissue remodeling. Macrophages play a dual role in this sequence of events since they contribute to tissue destruction, and also initiate fibroblast proliferation and collagen synthesis. In an inflammatory site, macrophages are found in close proximity to proliferating and activated fibroblasts facilitating the exchange of stimulatory and/or inhibitory signals (fig. 4A). Interestingly, multinucleated giant cells are also frequently associated with areas of fibrosis (fig. 4B), although their contribution to these events is currently unknown [35]. The mechanisms of tissue repair as manifested by collagen production and fibrosis continue to be intensely studied, as additional mononuclear cell-derived regulatory pathways are defined.

Early studies in silicosis suggested that silica-exposed macrophages released 'fibrogenic factors' which control collagen synthesis by fibroblasts [36]. The now classic studies of Ross and colleagues [37] further defined the role of the macrophage in mediating fibrotic events. Suppression of monocytes in an experimental model of wound healing by the administration of hydrocortisone and antimacrophage serum revealed pronounced retardation of the events central to fibrosis. Since those studies, many different monocyte-derived fibrogenic factors have been identified and characterized, and more recently, cloned and sequenced.

One of the first well-characterized macrophage-derived products shown to enhance fibroblast proliferation was IL-1 [38]. IL-1, which represents a family of related molecules with multiple biologic activities, stimulates fibroblast growth either directly or possibly, by increasing receptor number or affinity for other growth factors. IL-1 also stimulates fibroblasts to produce GM-CSF [28] which can, in turn, augment monocyte IL-1 production [39] providing a cycle of cellular activation and production of growth factors. In addition to proliferation, IL-1 increases matrix synthesis, type I and III collagens [40], at the transcriptional level, contributing to enhanced deposition of collagen in inflammatory sites. A functionally overlapping, but structurally distinct (only 3% amino acid homology) monocyte product, TNF, is also mitogenic for fibroblast monolayers in vitro and augments collagen synthesis [41], although fibroblasts have distinct receptors for IL-1 and TNF.

Another monocyte-derived growth factor important in the reparative phase is PDGF. Although originally defined as a product of the α-granules of platelets, PDGF has recently been identified as a secretory molecule of activated macrophages [42, 43]. A potent mitogen for fibroblasts, this monokine is also chemotactic for mesenchymal cells [44]. Its presence in early (platelets) and chronic (macrophages) stages of an inflammatory response may potentially contribute to physiologic and/or overabundant tissue repair. Similarly, FGF, with significant proliferative capacity for endothelial cells and fibroblasts [45], was characterized as a brain-derived molecule long before it was recognized as a monocyte polypeptide [46]. Synthesis of FGF which shares sequence homology with IL-1 [47] is regulated at the pretranslational level in activated macrophages [48]. One of the most effective inducers of FGF gene expression in monocytes is TGF-β, a 25,000 M_r homodimer released by platelets, and also secreted by macrophages upon activation [49]. Although mRNA for TGF-β is constitutively expressed by monocyte-macrophages, the cells require activation to secrete the polypeptide. Significantly, the secretion of TGF-β by activated macrophages, and the ability of TGF-β itself to activate macrophages [26], suggests an important autocrine and/or paracrine loop for regulation of monocyte-derived growth factors essential to fibroblast proliferation and tissue repair.

Associated with the accumulating population of fibroblasts is the observed deposition of connective tissue matrix (fig. 4). The accumulating fibroblasts are most likely stimulated by inflammatory cell-derived molecular signals to increase their synthetic activities with the generation of extracellular connective tissue matrix proteins. Interestingly, many of the polypeptide factors which regulate cell division also influence protein synthesis. TNF, IL-1 and TGF-β up-regulate the synthesis of fibrillar proteins (collagen) and of proteoglycans and glycoproteins (fibronectin, laminin and chondronectin) which constitute the ground substance of the extracellular matrix [50, review 51]. Although fibroblasts are interspersed among the inflammatory cells as the granuloma develops, matrix secretion is most apparent in the later stages of the inflammatory response. Encapsulation of the lesion within collagenous tissue helps to remove the granulomagenic agent from the circulation. Whereas the manifestations of normal tissue repair are represented by simple scar formation, in a chronic inflammatory response as exemplified in SCW granulomas, the prolonged release of growth-promoting and matrix-inducing factors may result in extensive collagen deposition with potentially detrimental consequences. Pathology due to chronic stimulation of fibrotic processes following tissue injury is apparent in granulomatous disorders such as tuberculosis, schistosomiasis and silicosis [1, 36, 52]. Mechanisms whereby mononuclear cells suppress matrix synthesis have also been identified [53, 54], although it is entirely unclear what determines continuation versus termination of the synthesis of collagen and other matrix proteins.

Matrix Remodeling

Matrix is continually turned over and remodeled by fibroblasts secreting collagenases, proteoglycanases, glycoproteases and other proteases. In inflammation, this process of remodeling may be heightened by the presence of molecular signals from inflammatory cells (PDGF, TNF, IL-1) which, in addition to their effects on growth and protein synthesis, also modulate the production of collagenase and PGE_2 [55, 56]. Remodeling and degradation of matrix components at the inflammatory site releases proteolytic fragments with biological importance. Collagen cleavage peptides are chemotactic for both monocytes and fibroblasts, as are the split products of elastin and fibronectin degradation [51]. Fibroblasts are responsible for generating enzymes capable of degrading matrix components, but monocytes also secrete collagenase [57], elastase, and other enzymes [21] critical to matrix degradation and the remodeling process. Of significance for the potential regulation of these degradative processes is the production of enzyme inhibitors. Activated fibroblasts and macrophages release a tissue inhibitor of metalloproteases (TIMP) which complexes with collagenase and other enzymes to inactivate them [58]. Although collagenase inhibitors are present, alterations in the homeostatic balance between catalytically active enzymes and inhibitors must contribute to the pathogenesis of fibrosis in many granulomatous diseases. The regulation of these enzymes and inhibitors may be important in determining the extent of tissue deposition and destruction, and ultimately, the outcome and reversibility of fibrosis.

Concluding Remarks

The monocyte-macrophages may be decisive in determining the chronicity of inflammatory lesions, the extent of tissue injury, pathways of tissue repair, the reversibility of fibrosis and residual pathologic manifestations. Since chronic inflammatory lesions can lead to tissue damage and/or extensive fibrosis, suppression of host granulomatous responses may reduce the pathologic consequences and possibly, increase survival. In view of the fact that many granulomatous responses require an interaction between antigen specific T cells and macrophages, various approaches have been attempted to suppress granulomatous lesions via immunosuppressive drugs and anti-inflammatory agents. Steroidal and nonsteroidal drugs dampen the response [13] and measures which suppress cell-mediated immune responses are often most efficient in inhibiting granuloma formation and its consequences [4]. Suppression of the granulomatous responses may also occur naturally with diminished lesion size evident in the later stages.

Although little is known of specific naturally occurring inhibitors of inflammatory cytokines, one possible moderating influence on these chronic lesions may be TGF-β [59]. TGF-β has recently been shown to be an extremely potent inhibitor of IL-1-dependent T cell proliferation [59, 60]. As it accumulates in inflamed lesions, TGF-β is activated from its latent form, interacts with receptors on T cells, and may down-regulate the T cell response to SCW and contribute to the spontaneous involution of the granulomas. Complex immune mechanisms are likely active in the spontaneous suppression of the granulomatous inflammatory response and other inhibitors are undoubtedly present. As we gain a better understanding of the signals that trigger the formation of granulomatous lesions as well as those that inhibit the response, we may be able to manipulate these signals by the use of selective immunotherapy and thereby alter the outcome of the response.

The ultimate goal of the host response to a foreign invader is the ingestion, containment, and/or elimination of the offending substance. Intended for the protective benefit of the host, a granulomatous inflammatory response can be detrimental when it induces exaggerated tissue destruction and repair with irreversible tissue pathology. Irrespective of the outcome, the key element in all granulomatous responses appears to be the extent and degree of macrophage activation.

Acknowledgments

The author is grateful to Dr. A. Hand for the electron micrographs, to Drs. E. Janoff, G. Bansal and J.B. Allen for helpful discussions and Sarah Smith for manuscript preparation.

References

1 Adams DO: The granulomatous inflammatory response. A review. Am J Pathol 1976:84:164–192.
2 Boros DL: Granulomatous inflammation. Prog Allergy. Basel, Karger, 1978, vol 24, pp 183–267.
3 Warren KS: A functional classification of granulomatous inflammation. Ann NY Acad Sci 1976;207:7–18.
4 Wahl SM, Allen JB, Dougherty S, et al: T lymphocyte dependent evolution of bacterial cell wall-induced hepatic granulomas. J Immunol 1986;137:2199–2209.
5 Wahl SM, Hunt DA, Allen JB, et al: Bacterial cell wall-induced hepatic granulomas. An in vivo model of T cell dependent fibrosis. J Exp Med 1986;163:884–902.
6 Ohanian SH, Schwab JH: Persistence of group A streptococci cell walls related to chronic inflammation of rabbit dermal connective tissue. J Exp Med 1967; 125:1137–1148.
7 Krause RM, McCarty M: Studies on the chemical structure of the streptococcal cell wall. I. The identification of a mucopeptide in the cell walls of groups A and A-variant streptococci. J Exp Med 1961;114:127–140.
8 Wilder RL, Allen JB, Wahl LM, et al: The pathogenesis of group A streptococcal cell wall induced polyarthritis in the rat: Comparative studies in arthritis resistant and susceptible inbred rat strains. Arthritis Rheum 1983;26:1442–1451.
9 Schwab JH, Allen JB, Anderle SK, et al: Relationship of complement to experimental arthritis induced in rats with streptococcal cell walls. Immunology 1982;46:83–88.
10 Weksler BB: Platelets; in Gallin JI, Goldstein IM, Snyderman R (eds): Inflammation: Basic Principles and Clinical Correlates. New York, Raven Press, 1988, pp 543–558.
11 Robey FA, Ohura K, Futaki S, et al: Proteolysis of human C-reactive protein produces peptides with potent immuno-modulating activity. J Biol Chem 1987; 262:7053–7057.
12 Kozin F, Cochrane CG: The contact activation system of plasma: biochemistry and pathophysiology; in Gallin JI, Goldstein IM, Snyderman R (eds): Inflammation: Basic Principles and Clinical Correlates. New York, Raven Press, 1988, pp 101–120.
13 Swisher J, Allen J, Feldman G, et al: Pharmacologic regulation of the development of streptococcal cell wall induced arthritis. J Leuk Biol 1988;44:305.
14 Shima K, Dannenberg AM, Ando M, et al: Macrophage accumulation, division, maturation and digestive and microbicidal capacities in tuberculous lesions. I. Studies involving their incorporation of tritiated thymidine and their content of lysosomal enzymes and bacilli. Am J Pathol 1972;67:159–180.

15 Spector WG, Lykke AWJ: The cellular evolution of inflammatory granulomata. J Path Bact 1966;92:163–177.
16 Poy M, Allen J, Wahl L, et al: Activation of mononuclear phagocytes by bacterial peptidoglycans: possible role in the pathogenesis of arthritis. J Leuk Biol 1988;44:296.
17 Skaleric U, Allen JB, Wahl SM: Superoxide release from rat macrophages exposed to bacterial cell wall fragments. J Dent Res, in press, 1988.
18 Dinarello CA: Biology of interleukin-1; in Sorg C (ed): Macrophage-Derived Cell Regulatory Factors. Cytokines. Basel, Karger, 1989, vol 1, pp 105–154.
19 Elias JA, Schreiber AD, Gustilo K, et al: Differential interleukin-1 elaboration by unfractionated and density fractionated human alveolar macrophages and blood monocytes: relationship to cell maturity. J Immunol 1986;135:3198–3204.
20 Wewers MD, Rennard SI, Hance AJ, et al: Normal human alveolar macrophages obtained by bronchoalveolar lavage have a limited capacity to release interleukin-1. J Clin Invest 1984;74:2208–2218.
21 Nathan CF: Secretory products of macrophages. J Clin Invest 1988;79:319–326.
22 Allen JB, Malone DG, Wahl SM, et al: The role of the thymus in streptococcal cell-wall induced arthritis and hepatic granuloma formation: Comparative studies of pathology and cell wall distribution in athymic and euthymic rats. J Clin Invest 1985;76:1042–1056.
23 Yocum DE, Allen JB, Wahl SM, et al: Inhibition by cyclosporin A of streptococcal cell wall induced arthritis and hepatic granulomas in rats. Arthritis Rheum 1986;29:262–273.
24 Warren KS, Domingo EO, Cowan RBT: Granuloma formation around schistosome eggs as a manifestation of delayed hypersensitivity. Am J Path 1967;51:735–748.
25 Doughty BL, Phillips SM: Delayed hypersensitivity granuloma formation and modulation around *Schistosoma mansoni* eggs in vitro. J Immunol 1982;128:37–42.
26 Wahl SM, Hunt DA, Wakefield L, et al: Transforming growth factor beta (TGF-β) induces monocyte chemotaxis and growth factor production. Proc Natl Acad Sci USA 1987;84:5788–5792.
27 Thorans B, Mermod JJ, Vassalli P: Phagocytosis and inflammatory stimuli induce GM-CSF mRNA in macrophages through posttranscriptional regulation. Cell 1987;48:671–679.
28 Zucali JR, Dinarello CA, Oblon DJ, et al: Interleukin-1 stimulates fibroblasts to produce granulocyte-macrophage colony-stimulating activity and prostaglandin E_2. J Clin Invest 1986;77:1857–1863.
29 Nathan CF, Prendergast TJ, Wiebe MR, et al: Activation of human macrophages. Comparison of other cytokines with interferon-gamma. J Exp Med 1984;160:600–605.
30 Smith PD, Lamerson CL, Wahl SM: Granulocyte-macrophage colony-stimulating factor augmentation of leukocyte effector cell function. J Cell Biochem, in press, 1988.
31 Wahl SM, McCartney-Francis N, Hunt DA, et al: Monocyte interleukin-2 receptor gene expression and interleukin-2 augmentation of microbicidal activity. J Immunol 1987;139:1342–1347.
32 McCartney-Francis N, Katona I. Mizel D, et al: Inducible gene expression for IL-2 receptors on human peripheral blood monocytes. Fed Proc 1987;46:1388.
33 Folkman J, Klagsbrun M: Angiogenic factors. Science 1987;235:442–447.

34 Wahl SM: Fibrosis: Bacterial cell wall induced hepatic granulomas; in Gallin JI, Goldstein IM, Snyderman R (eds): Inflammation: Basic Principles and Clinical Correlates. New York, Raven Press, 1988, pp 841–860.
35 Schlesinger L, Musson RA, Johnston RB Jr: Functional and biochemical studies of multinucleated giant cells derived from the culture of human monocytes. J Exp Med 1984;159:1289–1294.
36 Heppleston AG, Styles JA: Activity of a macrophage factor in collagen formation by silica. Nature 1967;214:521–523.
37 Leibovich SJ, Ross R: The role of the macrophage in wound repair: A study with hydrocortisone and antimacrophage serum. Am J Pathol 1975;78:71–91.
38 Schmidt JA, Mizel SB, Cohen D, et al: Interleukin-1, a potential regulator of fibroblast proliferation. J Immunol 1982;128:2177–2181.
39 Wong H, Hunt D, Dougherty S, et al: Colony-stimulating factor (CSF)-induced gene expression in human monocytes. FASEB J 1988;2:A875.
40 Canalis E, McCarthy T, Centrella M: Growth factors and the regulation of bone remodelling. J Clin Invest 1988;81:277–281.
41 Vilcek J, Palombella VJ, Henriksen-DeStefano D, et al: Fibroblast growth enhancing activity of tumor necrosis factor and its relationship to other polypeptide growth factors. J Exp Med 1986;163:632–643.
42 Shimokado K, Raines EW, Madtes DK, et al: A significant part of macrophage-derived growth factor consists of at least two forms of PDGF. Cell 1985;43:277–286.
43 Martinet Y, Bitterman PB, Mornex J, et al: Activated human monocytes express the c-sis proto-oncogene and release a mediator showing PDGF-like activity. Nature 1986;319:158–160.
44 Deuel TF, Senior RM, Huang JS, et al: Chemotaxis of monocytes and neutrophils to platelet derived growth factor. J Clin Invest 1982;69:1046–1049.
45 Thomas KA: Fibroblast growth factors. FASEB J 1987;1:434–440.
46 Baird A, Mormede P, Bohlen P: Immunoreactive fibroblast growth factor in cells of peritoneal exudate suggests its identity with macrophage-derived growth factor. Biochem Biophys Res Commun 1985;126:358–364.
47 Giminez-Gallego G, Rodkey J, Bennett C, et al: Brain-derived acidic fibroblast growth factor: complete amino acid sequence and homologies. Science 1985; 230:1385–1388.
48 Wahl SM, Wong H, McCartney-Francis N: Growth factors in inflammation and repair. J Cell Biochem, in press.
49 Assoian RK, Fleurdelys BE, Stevenson HC, et al: Expression and secretion of type β transforming growth factor by activated human macrophages. Proc Natl Acad Sci USA 1987;84:6020–6024.
50 Roberts AB, Sporn MB, Assoian RK, et al: Transforming growth factor type β: rapid induction of fibrosis and angiogenesis in vivo and stimulation of collagen formation in vitro. Proc Natl Acad Sci USA 1986;83:4167–4171.
51 Postlethwaite AE, Kang AH: Fibroblasts; in Gallin HI, Goldstein IM, Snyderman R (eds): Inflammation: Basic Principles and Clinical Correlates. New York, Raven Press, 1988, pp 577–598.
52 Wyler DJ, Wahl SM, Wahl LM: Hepatic fibrosis in schistosomiasis: egg granulomas secrete fibroblast stimulating factor in vitro. Science 1978;207:438–440.
53 Jimenez SA, Freundlich B, Rosenbloom J: Selective inhibition of human diploid fibroblast collagen synthesis by interferons. J Clin Invest 1984;74:1112–1116.

54 Korn JH, Halushka PV, LeRoy EC: Mononuclear cell modulation of connective tissue function. Suppression of fibroblast growth by stimulation of endogenous prostaglandin production. J Clin Invest 1980;65:543–554.
55 Dayer JM, Beutler B, Cerami A: Cachectin/tumor necrosis factor stimulates synovial cells and fibroblasts to produce collagenase and prostaglandin E_2. J Exp Med 1985;162:2163–2167.
56 Bauer EA, Cooper TW, Huang JS, et al: Stimulation of in vitro human skin collagenase expression by platelet-derived growth factor. Proc Natl Acad Sci USA 1985;82:4132–4136.
57 Wahl IM, Wahl SM, Mergenhagen SE, et al: Collagenase production by lymphokine activated macrophages. Science 1975;187:261–263.
58 Stricklin GP, Welgus HG: Human skin fibroblast collagenase inhibitor. J Biol Chem 1983;258:12252–12258.
59 Wahl SM, Hunt DA, Wong HL, et al: Transforming growth factor beta is a potent immunosuppressive agent which inhibits interleukin-1-dependent lymphocyte proliferation. J Immunol 1988;140:3026–3032.
60 Kehrl JH, Wakefield IM, Roberts AB, et al: The production of TGF-β by human T lymphocytes and its potential role in the regulation of T cell growth. J Exp Med 1986;163:1037–1050.

Dr. Sharon M. Wahl, Cellular Immunology Section, National Institute of Dental Research, National Institutes of Health, Bethesda, MD 20892 (USA)

Sorg C (ed): Macrophage-Derived Cell Regulatory Factors.
Cytokines. Basel, Karger, 1989, vol 1, pp 193–203

Monocyte and T-Cell-Activating Cytokines in the Acquired Immune Deficiency Syndrome (AIDS): Pathogenesis-Promoting or -Limiting Factors?

A. Billiau

Rega Institute, University of Leuven, Belgium

Endogenous cytokines are generally considered as important factors in the pathogenesis of various diseases, especially those which involve the immune system. Remarkably, in many instances they seem to be able to play a janus-like role, being disease-limiting under certain conditions and disease-promoting under others. Interferon-gamma (IFN-gamma) is a good case in point [3, 6, 23]. Here I wish to review and discuss evidence pertaining to a possible two-sided role of cytokines, including IFN-gamma, in the acquired immune deficiency syndrome (AIDS).

The Role of T-Cell Activation

Human immune deficiency virus (HIV), the agent which causes AIDS, is a persistent virus, which may be thought of as being exceptional because it primarily infects cells of the immune system, i.e. the system that is precisely responsible for keeping infectious agents out of the body. However, several other viruses, e.g. Epstein-Barr virus (EBV) and cytomegalovirus (CMV), also have their permanent residence in immunocytes: EBV in the B cell, CMV in the cells of the vessel wall. The most distinctive, if not the most important property of HIV is its ability to infect the T-helper (T_H lymphocytes. These cells, also designated as $T4^+$ cells, are recognized by the virus through their unique CD4 membrane protein.

Although there can be no doubt that invasion of T_H cells by HIV constitutes a crucial causal element in the depletion of this population of cells, it remains unclear why the onset of depletion is most often preceded by a latent phase, and why, once initiated, it still takes a long time to attain its full depth. Thus, in patients with AIDS, HIV RNA is undetectable in the majority of T cells [17], although it is possible that many more or possibly all $T4^+$ cells are carrying latent virus genomes. In vitro inoculation of a $T4^+$ cell population with HIV leads to infection of probably all cells, regardless of subset or state of activation. However, active replication of the virus seems to be determined by the activation/proliferation state of the infected cells [11, 24]. It thus appears that factors, other than the presence of T4 molecules, are necessary for viral replication to occur. That these factors may be cytokines is hinted at by the observation that HIV replication and HIV-induced CPE are readily observed in T_H cells that have been activated by exposure to lectins or antigen [11, 24].

Although the majority of T_H cells in AIDS patients appear to be uninfected, overall T-cell function is nevertheless impaired [for review, see 10, 25, 27]. In addition, B-lymphocyte function is also disturbed as is apparent from excess immunoglobulin (Ig) production, suggestive for polyclonal B cell stimulation [20]. Since both T- and B-cell function are under control of the cytokine cascade, their generalized impaired function also suggests involvement of cytokines in the pathogenesis of the disease.

The Role of Mononuclear Phagocytic Cell Activation

It is well known that normal functioning of T_H cells depends on their ability to engage in harmonious interactions with other immunocytes, in particular antigen-presenting cells. Among these, cells belonging to the mononuclear phagocyte (henceforth MPC) family have been identified as alternative hosts to the virus [15, 19], although only small numbers of CD4 molecules occur on their membranes. It has been suggested that MPCs, being long-lived cells, may harbor the virus in a dormant state, thus constituting a potential seed stock for infection of other cells [31, 37]. There is also evidence that active replication of the virus in MPC is stimulated by certain cytokines, in much the same way as infection of T_H cells preferentially occurs when these cells have first been activated. Thus, a cloned, chronically infected promonocyte cell line (U1) has been reported to increase its production of HIV after exposure to a cytokine mixture [12].

AIDS, a Disease Caused by Indiscriminate, Cytokine-Mediated Activation of MPCs and T Cells?

Activation of MPCs and T_H cells are both phenomena which, although still ill-defined mechanistically, are well known to be controlled by cytokines. The main T-cell-activating cytokine is interleukin-1 (IL-1) [8]; the main MPC-activating cytokine, on the other hand, is IFN-gamma [1, 21, 30, 36, 38]. Although both these cytokines are likely to be produced by a variety of cell types, it is generally accepted that IL-1 is a typical MPC product, while IFN-gamma is a typical T_H cell product. As a conseqence, by virtue of their ability to secrete these cytokines, MPCs and T_H cells can mutually activate each other. Under 'normal' conditions, i.e. in the absence of HIV, they seem to do so only when entering into close contact with each other as a result of recognition and binding by clonotypic T_H-cell receptors to antigenic epitopes presented by the MPC. In addition, it is implied that some negative feed mechanism exists which keeps the resulting mutual activation cycle in check. This well-controlled mutual activation of MPCs and T_H cells sets in motion a coordinated response of other immunocytes, an effect that, under normal conditions (i.e. in the absence of HIV), remains limited to those T and B cells that carry clonotypic membrane receptors which recognize the antigenic epitope(s) presented by the MPC. As a result, expanded clones of relevant antigen-specific T and B cells are generated.

On the basis of this generally accepted notion, one may try to predict what might happen if the MPC partner in the interaction happens to carry dormant HIV. In this case its incidental activation by the fact that it presents a foreign antigen to T cells may not only cause the release of IL-1 but also of HIV. Thus, in an HIV-carrying individual, each antigen-presenting event may result in infection of additional numbers of T_H cells.

At this point it may be helpful for our understanding of the role of cytokines in AIDS to consider the fact that an MPC which presents HIV antigens on its surface (whether it does so as a result of classical antigen uptake and partial degradation, or as a result of replication of endogenous HIV), has the ability for indiscriminate interaction with any T_H cell, the reason being that each T_H cell carries a receptor (the CD4 molecule) recognizing at least one antigen presented by the HIV-infected MPC (fig. 1). From this it would derive that, when an HIV-carrying individual is exposed to a foreign antigen, his antigen-presenting MPCs may engage in activation-generating interactions, not only with antigen-relevant T_H cells, but simply with all T_H cells. An important consequence could be that the antigen-specific

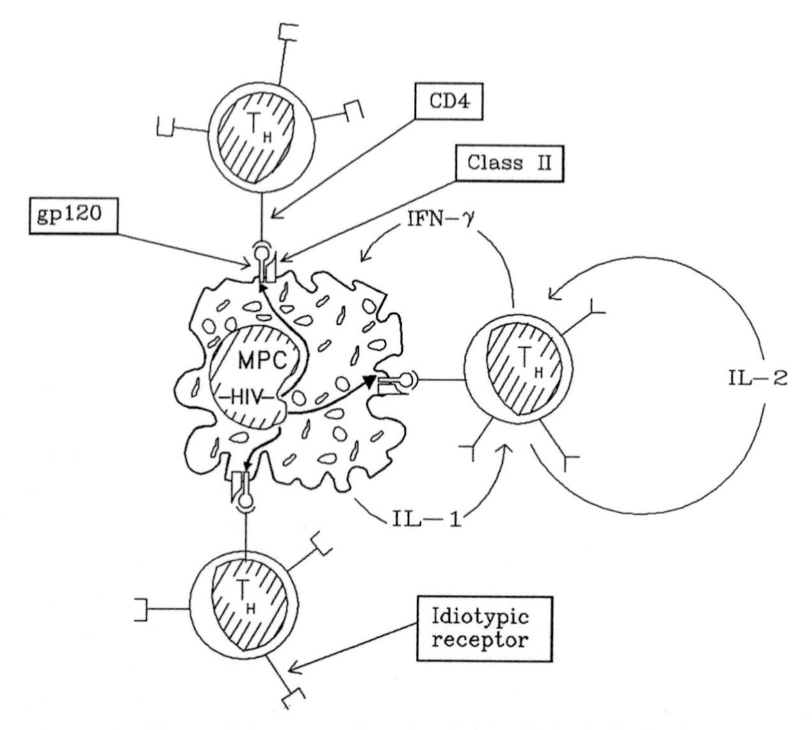

Fig. 1. A model mechanism for indiscriminate cytokine-mediated stimulation of T cells by activated HIV-infected mononuclear phagocytes (MPC). HIV-envelope protein (gp120) is presented on the surface of the infected MPC in association with Class II antigen and is therefore recognized by CD4-carrying T cells, irrespective of their idiotype specificity. This recognition triggers a self-perpetuating cytokine production loop.

interaction is diluted or even diverted, one of the possible reasons why the HIV carrier may be immune deficient. However, although not generating an adequate antigen-specific immune response, the interacting T_H cells may as yet produce their normal quotum of IFN-gamma. This IFN-gamma may in turn act back on naive bystander MPCs and convert them to additional HIV producers.

Indiscriminate activation of T_H cells, irrespective of their epitope specificity, may lead to equally indiscriminate activation of the corresponding B cells, a possible explanation for the presence, in AIDS patients, of an excess of nonspecific Ig.

Evidence for Excess Production of Cytokines in AIDS

Although involvement of cytokines in AIDS pathogenesis seems likely, solid evidence for it is scanty. AIDS patients, as well as patients with ARC, have been reported to have in their circulation an antiviral factor resembling IFN-alpha, but different from this IFN by the fact that its antiviral activity is not destroyed by acid [7, 9, 22]. The exact molecular nature, cellular origin and biological functions of this cytokine have not yet been elucidated. The same or a similar factor occurs in the serum of patients with certain autoimmune diseases, e.g. lupus erythematosus [33].

Blood mononuclear cells taken from patients at risk for developing AIDS have been reported to produce lesser amounts of IFN-gamma when stimulated in vitro than similar cells of control individuals [25, 27]. This could reflect the depletion of a population of T cells, e.g. the T_H population, necessary for IFN-gamma production. Alternatively, it could mean that T cells have lost some of their ability to produce IFN-gamma. It may even be taken as suggestive evidence for continuous stimulation of IFN-gamma production leading to hyporeactivity by an imbalance in production regulation [13]. It should also be noted that impairment of IFN-gamma producing ability has also been seen in other chronic diseases, e.g. multiple sclerosis [4, 40, 41] and lupus erythematosus [34]. In both these instances the observers have met with similar difficulties of interpretation. In the case of lupus the IFN-induced enzyme, 2′,5′-adenylate synthetase, was found to be elevated in mononuclear cells in vivo in both serum-positive and negative patients, suggesting that continuous low level production of IFN is an essential feature of the disease.

If HIV-infected individuals would produce IFN-gamma continuously or more frequently than healthy persons, then MPCs taken from such individuals might show signs of overexposure to IFN-gamma. In patients with clinically evident AIDS, monocyte function as tested by respiratory burst was found to be almost normal [32]. Other monocyte functions have been reported to be depressed. This was found to be the case with the chemotaxic response [39] as well as with the ability to assist T cells in their proliferative response [35]. However, in HIV-infected persons at risk for developing AIDS, various functions of both blood and lung alveolar mononuclear cells, e.g. phorbol ester-induced respiratory burst, sensitivity or resistance to intracellular microbial agents and response to IFN-gamma in vitro, have been found to be normal [26, 28].

The concept of an HIV-induced overproduction of IFN-gamma receives support from studies on the presence of neopterin in blood and urine of HIV-infected individuals [13, 14]. Neopterin is a not normally produced side product of the biosynthesis of the coenzyme tetrahydrobiopterin. Aside from being produced in patients suffering from a rare inborn error of metabolism, neopterin also occurs in blood and urine of cancer patients. In HIV-infected persons, levels of neopterin have been reported to increase in parallel with progression of disease, suggesting that a population of MPCs is under continuous strain.

Another indication that monocytes of HIV-infected individuals are oversolicited is the observation of increased spontaneous IL-1 production by such cells, together with a decreased responsiveness to additional stimulation [Folks et al., cf. 2].

Practical Implications of the Model

What actions does this model suggest for future research and for clinical management of HIV-infected patients? One aspect of the model is that it stresses the role of the MPC and its cytokine-mediated activation in the pathogenesis of AIDS. The early finding that the CD4 molecule of the T_H cell acts as a receptor for HIV entry has led to the oversimplified and perhaps partly erroneous current concept that immune deficiency in AIDS results from a direct viral attack on T_H cells, while, according to the proposed model, affinity of HIV for CD4 acts as a means for all T_H cells to be attracted to an HIV-infected and activated MPC. The model also suggests that AIDS may primarily not be an immune deficiency, but rather an autoimmune disease in the sense of a virus-driven *auto*nomous activation of the entire *immune* system. While such indiscriminate and sustained activation of all immunocytes may ultimately lead to deficiency by exhaustion, the primary deficiency may reside in the dilution and diversion of antigen-specific responses into a pool of aspecific T_H cell recruitment.

Since the model stresses the interaction between MPCs and T_H cells, and their mutual activation as crucial elements in the progression of the disease, it also heralds cytokines, such as IL-1 and IFN-gamma, as up-regulating factors in the disease.

One obvious implication is that one might be able to arrest the progression of AIDS if one could preemptively block the mortal activation cycle into which MPCs and T_H cells engage whenever a patient is exposed to a non-self antigen. In this context it would be extremely important to avoid such encounters with non-self, although the presence in the patient of other persistently infecting viruses such as CMV, EBV, VZV (varicella zoster virus), HSV (herpes simplex virus) or, for that matter HIV itself, may turn such endeavor into ridicule. One might also focus antiviral chemotherapy on episodes during which the patient is known to be exposed to foreign antigens.

The model certainly incites one to be cautious in contemplating immunostimulatory treatments of any kind. In fact, it rather inspires one to design protocols aiming at specifically blocking the production or action of particular cytokines in those HIV carriers who show signs of disease progression. The evidence reviewed here suggests that IL-1 and IFN-gamma may have to be considered as the prime candidates for such anticytokine therapy. Unfortunately there are some severe problems with this proposal. First, the oppposite approach, i.e. administration of IFN-gamma has been advocated on the grounds that production of this cytokine is impaired in HIV-infected individuals [25, 27], that MPCs taken from such persons develop a normally increased resistance to microbial agents after in vitro or in vivo treatment with IFN-gamma [25, 27] and that IFN-gamma can inhibit HIV replication in cells [29, 42, 43]. Obviously, one faces the well-known dilemma that cytokines, in particular IFN-gamma, may exert beneficial as well as detrimental influences on disease processes [3, 6, 23] and hence that the outcome of its administration or its blocking cannot reasonably be predicted.

Secondly, the choice of inhibitors of cytokines is at present very limited. However, from experiments in mice evidence has come forward that monoclonal antibodies against IFN-gamma can act in vivo to dramatically block certain undue effects of excessive endogenous MPC activation. Thus, the administration of neutralizing anti-IFN-gamma antibodies was found to protect mice against lethal shock caused by endotoxin [5]. Also, in a murine malaria model, the antibodies protected mice against lethal cerebral involvement [Grau et al., in preparation], a complication which is considered to be due to excessive release of TNF by activated intravascular MPCs in the brain [16].

It is noteworthy that mice given injections of anti-IFN-gamma could be kept for several months having detectable neutralizing antibody in the circulation and being protected against inflammatory challenge by endotoxin [18], but not showing signs of increased susceptibility to commonly occurring pathogens. Thus, fear that blocking of cytokine action in HIV carriers may further diminish their defense potential against infections may be groundless.

On the basis of these facts and considerations, one may advocate the use of anti-IFN-gamma antibodies (and anti-IL-1 antibodies, if available) to patients with clinically progressive HIV infection. The difficulty is, of course, that such antibodies, necessarily being of animal origin, would eventually become inactive because of the generation by the patient of human anti-murine Ig antibodies. The proposed therapy may also elicit the formation of anti-idiotype antibodies which may mimic the action of IFN-gamma and, according to the model, accelerate disease progression. Finally, one may object that the administration of heterologous Ig, since it constitutes an antigenic stimulus in itself, is contraindicated by the model.

In spite of these objections, there is hope that the anti-IFN-gamma antibodies may initially work as expected and may temporarily arrest progression of the disease or even revert its course. Even moderate beneficial effects of short duration may help to adjust our concepts of AIDS pathogenesis and hence to redefine the parameters of the therapeutical problem. Such outcome may also constitute a strong stimulus for the search for nonimmunogenic antagonists of cytokines, which may be the ultimate answer to AIDS.

Acknowledgements

Research in the author's laboratory is supported by grants from the Belgian Ministry of Science Policy (Concerted Research Actions), the Ministry of the Flemish Community, the ASLK/CGER Cancer Research Foundation and the Belgian National Fund for Scientific Research. The author appreciates the help of M. Van Ranst for literature research and of C. Callebaut for editorial help.

References

1 Arenzana-Seisdedos F , Virelizier JL, Fiers W: Interferons as macrophage activating factors. III. Preferential effects of interferon-γ on the interleukin-1 secretory potential of fresh or aged human monocytes. J Immunol 1985;134:2444–2448.
2 Barnes DM: Cytokines alter AIDS virus production. Science 1987;237:1627.
3 Billiau A: Interferons and inflammation. J Interferon Res 1987;7:559–567.
4 Billiau A, Carton H, Heirwegh K: A role for the interferon system in the pathogenesis of multiple sclerosis. J Biol Regul Homeostat Agents 1987;1:9–22.
5 Billiau A, Heremans H, Vandekerckhove F, et al: Anti-interferon-γ antibody protects mice against the generalized Shwartzman phenomenon. Eur J Immunol 1988; 17:1851–1854.

6 Brzoska J, Obert HJ: Interferon-gamma: ein janusköpfiger Mediator bei Entzündungen. Arzneimittelforschung 1987;37 suppl II:1410–1416.
7 Destefano E, Friedman RM, Friedman-Kien AE, et al: Acid-labile human leukocyte interferon in homosexual men with Kaposi's sarcoma and lymphadenopathy. J Infect Dis 1982;146:451–455.
8 Dinarello CA: Biology of interleukin-1. FASEB J 1988;2:108–115.
9 Eyster ME, Goedert JJ, Poon MC, et al: Acid-labile alpha interferon. A possible preclinical marker for the acquired immunodeficiency syndrome in hemophilia. N Engl J Immunol 1983;309:583–586.
10 Fauci AS: The human immunodeficiency virus: infectivity and mechanisms of pathogenesis. Science 1988:239:617–618.
11 Folks Y, Kelly J, Benn S, et al: Susceptibility of normal human lymphocytes to infection with HTLV-III/LAV. J Immunol 1986;136:4049–4053.
12 Folks TM, Justement J, Kinter A, et al: Cytokine-induced expression of HIV-1 in a chronically infected promonocyte cell line. Science 1987;238:800–802.
13 Fuchs D, Hausen A, Hengster P, et al: In vivo activation of $CD4^+$ cells in AIDS. Science 1987;235:356.
14 Fuchs D, Hausen A, Reibnegger G, et al: Neopterin as a marker for activated cell-mediated immunity. Immunol Today 1988;9:150–155.
15 Gartner S, Markovits P, Markovitz DM, et al: The role of mononuclear phagocytes in HTLV-III/LAV infection. Science 1986;233:215–219.
16 Grau GE, Fajardo LF, Piguet PF, et al: Tumor necrosis factor (cachectin) as an essential mediator in murine cerebral malaria. Science 1987;237:1210–1212.
17 Harper ME, Marselle LM, Gallo RC, et al: Detection of lymphocytes expressing human T-lymphotropic virus type III in lymph nodes and peripheral blood from infected individuals by in situ hybridization. Proc Natl Acad Sci USA 1986;83:772–776.
18 Heremans H, Dijkmans R, Sobis H, et al: Regulation by interferons of the local inflammatory response to bacterial lipopolysaccharide. J Immunol 1987;138:4175–4179.
19 Ho DD, Rota TR, Hirsch M: Infection of monocyte/macrophages by human T lymphotropic virus. H Clin Invest 1986;77:1712–1715.
20 Lane CH, Masur H, Edgar LC, et al: Abnormalities of B-cell activation and immunoregulation in patients with the acquired immunodeficiency syndrome. N Engl J Med 1983;309:453–458.
21 Le J, Vilcek J: Lymphokine-mediated activation of monocytes: neutralization by monoclonal antibody to interferon-γ. Cell Immunol 1984;85:278–283.
22 Levin S, Hahn T, Handzel ZT, et al: Activated interferon system in healthy homosexual men. Antiviral Res 1985;5:229–240.
23 Marx JL: Cytokines are two-edged swords in disease. Science 1988;239:257–258.
24 McDougal JS, Mawle A, Cort SP, et al: Cellular tropism of the human retrovirus HTLV-III/LAV I. Role of T-cell activation and expression of the 4 antigen. J Immunol 1985;135:3151–3162.
25 Murray HW, Rubin BY, Masur H, et al: Impaired production of lymphokines and immune (gamma) interferon in the acquired immunodeficiency syndrome. N Engl J Med 1984;310:883–888.
26 Murray HW, Gellene RA, Libby DM, et al: Activation of tissue macrophages from

AIDS patients: in vitro response of AIDS alveolar macrophages to lymphokines and interferon-γ. J Immunol 1985;135:2374–2377.

27 Murray HW, Hillman JK, Rubin BY, et al: Patients at risk for AIDS-related opportunistic infections: clinical manifestations and impaired gamma interferon production. N Engl J Med 1985;313:1504–1510.

28 Murray HW, Scavuzzo D, Jacobs JL, et al: In vitro and in vivo activation of human mononuclear phagocytes by interferon-γ. J Immunol 1987;138:2457–2462.

29 Nakashima H, Yoshida T, Yamamoto N: Recombinant human interferon gamma suppresses HTLV-III replication in vitro. Int J Cancer 1986;38:433–436.

30 Nathan CF, Murray HW, Wiebe ME, et al: Identification of interferon-γ as the lymphokine that activates human macrophage oxidative metabolism and antimicrobial activity. J Exp Med 1983;158:670–689.

31 Nicholson JK, Cross GD, Callaway CS, et al: In vitro infection of human monocytes with human T lymphotropic virus type III/lymphadenopathy-associated virus (HTLV-III/LAV). J Immunol 1986;137:323–329.

32 Pennington JE, Groopman JE, Small GJ, et al: Effect of intravenous recombinant γ-interferon on respiratory burst of blood monocytes from patients with AIDS. J Infect Dis 1986;153:609–612.

33 Preble OT, Black RJ, Friedman RM, et al: Systemic lupus erythromatosus: presence in human serum of an unusual acid-labile leukocyte interferon. Science 1982; 216:429–431.

34 Preble OT, Rothko K, Klippel JH, et al: Interferon-induced 2′,5′-adenylate synthetase in vivo and interferon production in vitro by lymphocytes from systemic lupus erythematosus patients with and without circulating interferon. J Exp Med 1983;157:2140–2146.

35 Prince HE, Moody DJ, Shubin BI, et al: Defective monocyte function in acquired immune deficiency syndrome (AIDS): evidence from a monocyte-dependent T-cell proliferative system. J Clin Immunol 1985;5:21–25.

36 Robert WK, Vasil A: Evidence for the identity of murine gamma interferon and macrophage activating factor. J Interferon Res 1982;2:519–532.

37 Salahuddin SZ, Rose RM, Groopman JE, et al: Human T lymphotropic virus type III infection of human alveolar macrophages. Blood 1986;68:281–284.

38 Schreiber DR, Pace JL, Russell SW, et al: Macrophage-activating factor produced by a T cell hybridoma: physicochemical and biosynthetic resemblance to γ-interferon. J Immunol 1983;131:826–832.

39 Smith PD, Ohura K, Masur H, et al: Monocyte function in the acquired immune deficiency syndrome — defective chemotaxis. J Clin Invest 1984;74:2121–2128.

40 Vervliet G, Claeys H, Vanhaver H, et al: Interferon production and natural killer (NK) activity in leukocyte cultures from multiple sclerosis patients. J Neurol Sci 1983;60:137–150.

41 Vervliet G, Carton H, Meulepas E, et al: Interferon production by cultured peripheral leucocytes of MS patients. Clin Exp Immunol 1984;58:116–126.

42 Wong GHW, Krowka JF, Stites DP, et al: In vitro anti-human immunodeficiency virus activities of tumor necrosis factor-α and interferon-γ. J Immunol 1988; 140:120–124.

43 Yamamoto JK, Barré-Sinoussi F, Bolton V, et al: Human alpha- and beta-interferon but not gamma- suppress the in vitro replication of LAV, HTLV-III, and ARV-2. J Interferon Res 1986;6:143–152.

Note Added in Proof

While this paper was in press, a recently published article (August, 1988) came to our attention, entitled 'AIDS as Immune System Activation: A Model for Pathogenesis', by M.S. Ascher and H.W. Sheppard [Clin Exp Immunol 1988;73:165–167]. In this paper the authors enounce the same hypothesis as that proposed independently by us. In particular, the following statement is made: 'In contrast to the obvious role of this molecule (CD4) in facilitating infection of T-cells, we feel that HIV infection of antigen presenting cells results in inappropriate activation of T-cell proliferation and mediator release.'

Prof. A. Billiau, Laboratory of Immunobiology, Katholieke Universiteit Leuven,
Rega Instituut, Minderbroedersstraat 10, B–3000 Leuven (Belgium)

Sorg C (ed): Macrophage-Derived Cell Regulatory Factors.
Cytokines. Basel, Karger, 1989, vol 1, pp 204–228

Fibroblast Growth Factor

Peter Böhlen

American Cyanamid Company, Medical Research Division,
Pearl River, N.Y. (USA)

Introduction

Macrophages and monocytes synthesize and secrete growth-promoting activities for a host of different cell types. Frequently, such activities have been termed macrophage-derived growth factors (MDGFs). However, this term is ambiguous because different investigators have used it to describe mitogenic activities for different target cells. Although some of the described MDGFs may represent structurally identical or related substances, it is now established that macrophages produce and secrete a variety of different growth factors. This chapter reviews in a first part the present knowledge of fibroblast growth factor (FGF), a well-characterized growth factor present in macrophages and, more importantly, in many tissues. In the second part, an attempt is made to bring what is generally known about macrophage-derived fibroblast growth activities into the context of known growth factors.

Fibroblast Growth Factor[1]

Chemistry and Molecular Biology

While tissue extracts were already known in the early 1940s to stimulate the proliferation of cultured fibroblasts and proteins with this capacity were partially purified and named FGF in the 1970s [4, 5], it was not until 1983 when acidic and basic mitogens were purified to homogeneity from brain (acidic FGF, aFGF [6]) and pituitary (basic FGF, bFGF [7]). The complete

[1] The topic of FGF has been reviewed previously [1–3].

amino acid sequences of aFGF [8] and bFGF [9] have been established. Both mitogens are single-chain proteins with molecular weights ranging from 15 to 18 kd. This size heterogeneity arises from N-terminal truncation of the mitogens, presumably as a consequence of the action of endogenous proteases liberated during tissue extraction under certain conditions [10–18]. The various N-terminally truncated FGFs seem to be biologically fully active. The sequences of aFGF and bFGF are related (55% homology), suggesting evolution of the two mitogens from a common ancestral gene. Furthermore, bFGFs from different species (and to a lesser degree also aFGFs) are extremely conserved [8, 9, 19–21], indicating possibly strong evolutionary pressure to conserve important biological function.

cDNAs encoding human aFGF and bovine bFGF have been cloned [22, 23] and nucleotide sequence and genomic organization of the human bFGF gene are known [20]. The human bFGF and aFGF genes are single-copy genes which are located on chromosomes 4 [24] and 5 [20], respectively. The cDNA derived from aFGF mRNA possesses an open reading frame that begins with a methionine initiator triplet and codes for a 155-residue protein. With the exception of the deleted N-terminal methionine residue, the sequence translated from the open reading frame of the aFGF mRNA corresponds exactly to the longest (untruncated) isolated form of the mitogen consisting of 154 amino acid residues [16]. Some evidence suggests that the bFGF mRNA has a similar structure. Untruncated bFGF is also a 154-residue protein [14] and the bFGF mRNA contains a methionine codon (presumably the translation initiation codon) at precisely the same location as is found in the mRNA of aFGF. However, in the bFGF mRNA the putative initiation codon is not immediately preceded by a termination codon and the mRNA possesses an 5′-extended open reading frame of undetermined length [20, 23]. In view of this uncertainty and of some experimental evidence that a longer form of bFGF might indeed exist [25], the length of the coding sequence of the bFGF mRNA is still debated.

The question regarding the translation initiation site of the FGF mRNAs is important for an understanding of the biosynthesis and secretion of the FGFs. Since the aFGF mRNA coding sequence and the amino acid sequence of the isolated mitogen are identical with the exception of the N-terminus (the protein lacks the N-terminal methionine residue and is acetylated [16]), one must conclude that the biosynthesis of this protein does not involve the usual initial synthesis of a larger precursor protein that is subsequently cleaved to yield the bioactive mature mitogen. Furthermore, the same evidence indicates that the aFGF protein is synthesized without a typical N-terminal signal sequence mediating protein export from the cell. In this context, it is presently unknown how aFGF is secreted from mitogen-producing cells. The biosynthetic pathway of bFGF is less well understood but in many respects it seems to closely resemble that of aFGF (154-residue mature protein, processing of the N-terminus by removal of methionine and acylation [26], lack of a typical N-terminal signal peptide sequence). There is also clear evidence that bFGF, despite the apparent absence of a signal peptide, is secreted from cells [27], presumably by an unknown mechanism. However, given the present uncertainty with respect to the location of translation initiation on the bFGF mRNA, the possibility cannot be excluded that bFGF is synthesized via a larger precursor which could contain a normal signal sequence. The FGF mRNA levels in cells are very low, possibly because of mRNA instability [22, 23]. However, tissue levels of FGF protein are relatively high [28, 29] suggesting that constitutively released FGF may accumulate outside the cell in the extracellular matrix (see below).

Distribution in Tissues and Cells

Mitogens structurally identical or closely similar to one of the several characterized forms of bFGF have been isolated from many normal and tumor tissues [2]. The distribution of aFGF was originally thought to be limited to neural tissues such as brain and retina [14, 30], but more recently this mitogen has also been found in other tissues such as kidney, myocardium, and bone [17, 31, 32]. The wide tissue distribution of bFGF is a consequence of the protein being synthesized by a great many cell types, including vascular endothelial cells [33], corneal endothelial and lens epithelial cells [34], fibroblasts [35], adrenal cortex [36] and granulosa [37] cells, and macrophages [38]. Acidic FGF is also synthesized in a variety of cell types such as smooth muscle cells [39] and neurons [40]. Furthermore, many tumor cells synthesize one or the other of these growth factors [35, 40–45].

An important issue relates to the question as to whether FGFs are present in serum or in other physiological fluids. FGF has not been detected unequivocally in serum so far and the favored view is that FGFs are not circulating hormones, but rather locally active paracrine or autocrine tissue factors [3]. Some data suggest, however, that aFGF and/or bFGF may occur in low concentrations in certain other biological fluids such as vitreous [46, 47], synovial fluid [48], urine [49, 50] and glioma cyst fluid [Spinas and Böhlen, unpubl.].

With the structural characterization of the FGFs it has become clear that a variety of growth factors once thought to represent unrelated mitogens are identical to one or the other of the FGFs. They include endothelial cell growth factor (ECGF), retina-derived growth factor (RDGF), eye-derived growth factor (EDGF), brain-derived growth factor (BDGF), cartilage-derived growth factor, astroglial growth factor (AGF), and prostatropin. In addition, FGFs are sometimes referred to as heparin-binding growth factors, in reference to their high affinity for the polysulfated glycosaminoglycan heparin [29, 51].

Secretion of FGF and Interaction with the Extracellular Matrix

Although typical N-terminal signal peptide sequences mediating protein export seem to be absent in the FGFs, these mitogens are nevertheless secreted from cells. Interestingly, in cultured cells, bFGF is secreted into the extracellular matrix produced by the cells rather than into the medium [52] and in certain tissues, bFGF has been localized in the basement membrane [53]. Given the strong affinity of FGF for heparin-like structures, it is likely that the mitogens bind to heparan sulfate proteoglycans which are abundant basement membrane constituents. Indeed, FGF is released into the culture medium when the matrix produced by cultured endothelial cells is treated by a heparan sulfate-degrading enzyme [54]. It is intriguing to speculate that the apparently directed secretion of FGF from cells into their basement membranes may constitute a mechanism by which cells could accumulate and store large amounts of mitogen outside the cell. This pool of bound and presumably inactive FGF would be readily available upon activation by signals that cause elevated levels of certain matrix-degrading enzymes. Such a mechanism could be beneficial in a variety of physiological situations of which the biological response to tissue injury is a prototype (see below).

Target Cells

Many cells, mostly of mesenchymal and neuroectodermal origin, respond to FGF[2]. FGF is mitogenic for most but not all of these cells. In many cell types, FGF also induces nonmitogenic reactions. These will be discussed later in the context of possible biological functions of FGF.

[2] Because aFGF and bFGF possess qualitatively indistinguishable biological properties in most systems investigated, the general term FGF is used frequently. Whenever there is evidence for qualitative differences in the biological activities of aFGF and bFGF, this fact will be mentioned.

The FGF Receptor. Membrane proteins that bind FGFs with high affinity and high specificity have been identified on several cell types including baby hamster kidney (BHK) cells [55], endothelial cells [56], smooth muscle cells [39], fibroblasts [57], myoblasts [57], epithelial lens cells [58], and PC12 pheochromocytoma [59], and a variety of other tumor cells [60].

The FGF receptor has not yet been cloned. It appears to be a single-chain polypeptide with a molecular weight ranging from 110 to 150 kd. The receptors bind FGFs with high affinity (kd = 10–80 p*M*) with receptor numbers ranging from 2,000 to 80,000 per cell. Moscatelli [61] distinguished between specific high-affinity FGF-binding sites and less specific low-affinity sites composed of heparin-like structures. Apparently, only the high-affinity binding sites are capable of mediating the FGF signal.

Consistent with the finding that the FGF receptor appears to bind bFGF and aFGF, with few exceptions cells respond to both FGFs in a qualitatively indistinguishable manner. bFGF is usually 10- to 100-fold more potent than aFGF [2, 62], likely reflecting a higher affinity of bFGF for the FGF receptor.

Many cells that respond to FGF also synthesize this growth factor (fibroblasts, vascular and corneal endothelial cells, smooth muscle cells, granulosa cells, adrenal cortex cells, and neurons). Such cells, if capable of releasing FGF, might stimulate themselves in an autocrine fashion but it is still unknown whether this type of regulation is physiologically relevant. Nonetheless, this concept has received considerable attention in attempts to explain uncontrolled cellular proliferation in tumors, in particular because a number of tumor cells are known to synthesize bFGF and to respond to it (glioma [45, 63], rhabdomyosarcoma [43], leukemia [35], hepatoma cells [64]). It was also suggested that bFGF-like activity secreted from melanoma cells may contribute to the malignant phenotype of melanocytes which possess bFGF receptors [65].

Signal Transduction. The binding of FGF to its receptor induces several of the common growth factor-related signal-transducing events. FGF causes the rapid expression of the fos and myc proto-oncogenes [66–68]. Several potential pathways have been implicated in the signaling mechanisms of various cell types, i.e. cyclic nucleotide/protein kinase A activation [67, 69], diacylglycerol/protein kinase C activation [67, 70, 71], calcium influx [68, 70], and receptor tyrosine kinase activation [72]. Interestingly, FGF-induced diacylglycerol production does not seem to be associated with inositol trisphosphate generation [68, 69, 73, 74] but rather, at least in one system, with the production of inositol monophosphate [74] suggesting the possibility of phosphatidylinositide metabolism by phospholipases other than phospholipase C [68, 75]. FGF also induces the phosphorylation of the S6 ribosomal protein by S6 kinase [68, 76], as well as a variety of other cytosolic proteins [75, 77]. Additional mechanisms may play a part in the transduction of the growth factor signal, e.g. the activation of the Na/H antiport [68].

Other Cellular Responses. FGF induces in cultured cells a variety of additional responses which may not readily be classified but seem to stage the cell to perform more specific functions. Typical such responses include a reversible change of morphology [2] brought about by cytoskeletal rearrangement [2], and the induction of the pleiotypic response which includes the stimulation of cellular transport systems, polyribosome formation, activation of ornithine decarboxylase, and RNA, DNA and protein synthesis [2, 77]. The growth factor also induces proteins of unknown function, some of which are major secreted proteins [78], and cells treated with FGF can assume a 'transformed' phenotype associated with anchorage-independent proliferation [2, 79]. In a variety of cells, FGF delays the onset of senescence [2].

Structure-Activity Relationships

FGF-Receptor Interaction. At present little is known about the structural elements which convey biological activity of the FGFs. Synthetic bFGF peptide fragments have been used to identify potential receptor and heparin-binding areas [80, 81]. Peptides corresponding to sequence locations near the N-terminus (bFGF (24–68)) and the C-terminus (bFGF (106–115)) were found to bind heparin and to displace labeled bFGF from its receptor. Furthermore, at high concentrations those peptides were able to interfere with the biological activity of bFGF in certain test systems or mimic it in the absence of bFGF, and thus are partial receptor (ant)agonists. The data suggest that receptor binding involves the interaction of two separate FGF sequence domains with the receptor. It is questionable whether a disulfide bond in the FGFs contributes to proper protein folding for binding because alteration of cysteines in aFGF is not essential for biological activity [82, 83]. It is possible that heparin or related glycosaminoglycans contribute to the stability of the FGF tertiary structure. FGFs, which bind heparin with particularly high affinity, are thought to possess two heparin-binding sequences [9]. Furthermore, heparin protects FGFs from inactivation [84] and potentiates the biological activity of aFGF in vitro and in vivo [2, 3].

Homology to Other Proteins. Highly interesting sequence homologies were recently discovered to exist between FGFs and two cellular oncogene products (table 1). The first is the product of the int-2 oncogene [85] which has been implicated in the virally induced formation of mouse mammary tumors but also appears to be expressed normally in a very restricted manner in early mouse embryogenesis. The second is a protein which is encoded by the hst oncogene isolated from Kaposi's sarcoma [86] and a human stomach tumor [87]. The hst protein is 39 and 32% homologous to bFGF and the int-2 protein, respectively. The biological activities of the int-2 and hst oncoproteins have not been studied in detail yet and it is unknown to what extent FGF, int-2 and hst proteins are related biologically. On the basis of the possible biological functions of FGF (see below) it will be of obvious interest to study the relative roles of these new members of FGF family in normal development and tumor formation.

Biological Activities of FGF

While FGF was originally discovered on the basis of its mitogenic activity for cultured fibroblasts, it is now evident that this molecule acts as a mitogen on many diverse cells of mostly mesenchymal or neuroectodermal origin. Equally important, FGF possesses activity as a modifier of a multitude of nonmitogenic cell functions which include chemotactic activity, induction or suppression of cell-specific protein synthesis or secretion, regulation of cellular differentiated functions, and modulation of endocrine and neural functions. FGF is a very potent agent: in most biological test systems bFGF is active at concentrations in the low picomolar range while aFGF in most systems is 10- to 100-fold less potent. At first glance the multitude of activities of the FGFs may appear bewildering and difficult to interpret. However, with the data now at hand, many of those activities can be seen in the context of certain physiological functions.

Vascularization of Tissues. FGFs induce new capillary blood vessel growth in various pharmacological animal models [9, 11, 12, 88]. Capillary formation is a complex and poorly understood process, of which important elements are (a) endothelial cell proliferation, (b) the sprouting of new capillaries from existing ones (requiring migration of endothelial cells), and (c) the breakdown of extracellular matrix surrounding existing capillaries (providing space for ingrowth of new blood vessels). Physiological angiogenesis factors are thought to mediate many, if not all of those activities. FGF possesses these properties. It stimulates the proliferation of vascular endo-

Table 1. Sequence homology between FGF, int-2, and hst proteins

```
haFGF                                                            MAEGEITTF----TAL
hbFGF                                                            MAAGSITTLPALPEDG
hst       MSGPGTAAVALLPAVLLALLAPWAGRGGAAAPTAPNGTLEAELERRWESLVALSLARLPVAAQPKEAAV
int-2     MGLIWLLLLSLLEPSWPTTGPGTRLRRD------------------------------------------AGG

haFGF     TEKFNLPPGNYKKPKLLYCSNG-GHFLRILPDGTVDGTRDRSDQHIQLQLSAESVGEVYIKSTETGQYLAMDT
hbFGF     GSGA-FPPGHFKDPKRLYCKNG-GFFLRIHPDGRVDGVREKSDPHIKLQLQAEERGVVSIKGVCANRYLAMKE
hst       QSGAGDYLLGIKRLRRLYCNVGIGFHLQALPDGRIGGAHA-DTRDSLLELSPVERGVVSIFGVASRFFVAMSS
int-2     RGGVYEHLGGAPRRRKLYCATK--YHLQLHPSGRVNGSLE-NSAYSILEITAVEVGVVAIKGLFSGRYLAMNK

haFGF     DGLLYGSQTPNEECLFLERLEENHYNTYISKKH--------------AEKNWFVGLKKNGSCKRG--PRTHYG
hbFGF     DGRLLASKCVTDECFFFERLESNNYNTYRSRKY----------------TSWYVALKRTGQYKLG--SKTGPG
hst       KGKLYGSPFFTDECTFKEILLPNNYNAYESYKY----------------PGMFIALSKNGKTKKG--NRVSPT
int-2     RGRLYASDHYNAECEFVERIHELGYNTYASRLYRTGSSGPGAQRQPGAQRPWYVSVNGKGRPRRGFKTRR--T

haFGF     QKAILFLPLPVSSD
hbFGF     QKAILFLPMSAKS
hst       MKVTHFLPRL
int-2     QKSSLFLPRVLGHKDHEMVRLLQSSQPRAPGEGSQPRQRRQKKQSPGDHGKMETLSTRATPSTQLHTGGLAVA
```

thelial cells in vitro [7, 28, 89, 90] and in vivo [91]. FGF also stimulates the migration of endothelial cells [64, 92] and the release from endothelial cells of the matrix-degrading proteases collagenase [64] and plasminogen activator [61, 64]. FGF induces a variety of additional in vitro effects in endothelial cells which may support the process of neovascularization. These include (a) the maintenance of a properly differentiated state in cultured endothelial cells (expression of a nonthrombogenic apical surface [93], preferential synthesis of certain types of collagen [94], delayed cell senescence [2], attachment to the substratum [81]); (b) the rearrangement of endothelial cells into tubular structures resembling blood capillaries (angiogenesis in vitro) [95, 96], and (c) the proliferation of smooth muscle cells [62] and of pericytes [91], both of which are associated with new blood vessels.

Neovascularization is clearly a tightly regulated process. While endothelial cell turnover in blood vessels under normal circumstances is very slow (in the order of years) and angiogenesis is virtually absent in the healthy adult organism, it is temporarily induced in certain physiological conditions (wound healing, menstrual cycle, pregnancy, embryogenesis, organ growth). Rampant unregulated neovascularization is a characteristic of the growth of most solid tumors. The question as to whether FGF is involved in the physiological or pathophysiological regulation of new blood vessel growth is obviously of great interest. Some evidence supports this notion. First, bFGF is present in relatively high quantity in probably all vascularized tissues, including tumor tissues. Second, in some embryonic tissues FGF is expressed at the time when vascularization of these tissues begins [97–99]. Third, neutralizing anti-bFGF antibodies administered to chondrosarcoma-bearing mice were found to significantly inhibit the growth of this tumor [100], suggesting the possibility of tumor growth inhibition via interference with FGF-dependent tumor vascularization.

The potential role of FGF in the regulation of vascularization is also apparent in the development and maturation of the ovarian corpus luteum in which a vascular network develops around the follicle and invades the previously avascular granulosa cell layer, presumably to facilitate luteinization and for nourishment of the developing corpus luteum [101]. Angiogenesis and endothelial cell mitogenic factors found in corpus luteum, follicles and follicular fluid [102–105] are thought to be responsible for the hormone-induced neovascularization. At least some of these factors are now identified as bFGF [11] and more recently it was shown that the granulosa cells, the target of this neovascularization, synthesize bFGF [37] and thus may be responsible for capillary blood vessel attraction. The vascularization of other tissues, including tumors, is thought to involve similar mechanisms whereby the tissue to be vascularized produces FGF or other signals that activate FGF stored in the extracellular matrix which in turn attract capillary blood vessels. In this context, it should be remembered that other factors such as transforming growth factors alpha (TGF-alpha) [106] and beta (TGF-beta) [107], tumor necrosis factor (TNF) [108], angiogenin [109], and prostaglandin E_2 [110] also possess angiogenic properties. However, their mechanisms of action and potential functions in the physiological regulation of neovascularization remain largely unknown.

Tissue Development and Differentiation. Recent data imply a role for FGF in early embryonic development that is broader than the above-described involvement of the growth factor in the vascularization of embryo tissues. In the fertilized Xenopus egg, exogenous FGF, unlike other known growth factors, mimics the effect of the ventrovegetal signal responsible for the differentiation of ectoderm into mesodermal structures [111]. Since the bFGF gene is also transcribed in the Xenopus egg [112], this evidence suggests that FGF may be a physiological mesoderm-inducing factor in the Xenopus embryo. Furthermore, bFGF mRNA is expressed in a variety of mouse embryo tissues [20] which further supports the notion that this growth factor may be involved in development.

Considerable evidence obtained from cell culture experiments points to a role of FGF in the induction or maintenance of a properly differentiated state of cells in tissues. For example, FGF promotes the differentiation of preadipocyte fibroblasts into adipocytes [113] and induces in PC12 pheochromocytoma cell line the expression of characteristics which resemble those typically seen in neurons [114]. Under FGF influence, chondrocytes express a different set of glycosaminoglycans than in the absence of mitogen [115] and epithelial lens cells differentiate into crystallin-producing lens fiber cells [116]. In muscle cells, FGF induces the synthesis of muscle-specific myosin [117] but interestingly, it inhibits other differentiated functions such as the expression of creatine kinase [118] and alpha actin [119]. It should be noted that in many cell types FGF is capable of stimulating proliferation as well as differentiation, at least in vitro. How these apparently contrasting functions are regulated in vivo is unclear.

Wound Healing/Tissue Regeneration. Tissue repair is a highly coordinated process that integrates several key functions such as the mounting of an appropriate immune response, the degradation of injured tissue, and the formation of new tissue and its vascularization. FGF has been shown to promote the overall healing process in experimental skin [91, 120], eye [121–124], and joint [125] injuries. At the basis of the regenerative properties of FGF might be the wide distribution of the mitogen in organs which potentially could need repair (many cell types making up those tissues synthesize FGF) [2]. In certain tissues FGF has been localized to the basement membranes [53] and evidence is accumulating which suggests that FGF is likely to be present in the extracellular matrix of other tissues as a heparan sulfate-bound storage form.

Other important aspects are that many cell types of the damaged tissues can respond to FGF by proliferation (e.g. fibroblasts, vascular endothelial cells, pericytes [91]. smooth muscle cells [62], chondrocytes [62], osteoblasts [126], corneal epithelial and endothelial cells [62, 123], keratinocytes [127]) and that many of those cells also respond to FGF by synthesizing and secreting proteins and glycosaminoglycans needed for tissue repair. For example, FGF stimulates in some of those cells the release of matrix-degrading enzymes such as collagenase and plasminogen activator [128, 129], the synthesis of matrix components such as collagen, fibronectin, and proteoglycans [91, 120, 130, 131], and an inhibitor of matrix degradation, plasminogen activator inhibitor [129]. Finally, FGF is also a chemoattractant for various cells which can contribute to injury repair [131]. Thus, FGF stimulates cellular responses in many of the events that are considered essential for the progression of tissue repair.

FGF has also been shown to support tissue regeneration in various models. It stimulates the morphological and functional regeneration of adrenal cortex tissue [132] after incomplete ablation or transplantation of adrenal fragments to other sites. The presence of FGF in the adrenal [12], the synthesis of the mitogen in adrenocortical cells [36], the mitogenic response of those cells towards FGF [62], and the angiogenic and probably other properties (e.g. matrix synthesis) of FGF are likely at the basis of this regenerative effect. Furthermore, FGF induces regrowth of amputated frog limbs [133] and stimulates blastema formation in denervated newt limb [134], a tissue well known for its regenerative properties. Finally, FGF has also been reported to promote the regeneration of the newt lens in organ culture [135]. Again, some known in vitro activities of FGF, i.e. the mitogenic stimulation of blastema [136] or epithelial lens cells [137], or the induction of the synthesis of the lens-specific protein crystallin [116], are consistent with the in vivo regenerative properties of FGF.

Modulation of Neural Function. FGF has profound effects on the principal neural cells. It is mitogenic for glial cells [138, 139] and stimulates several of their nonmitogenic functions, such as migration [140], morphological maturation [141], expression of the glial-specific proteins GFAP, glutamine synthetase, and S100 protein [141]. FGF is also active on neurons. While it is not mitogenic for mature neuronal cells, it stimulates the proliferation of their precursor cells, the neuroblasts [142] and it induces differentiated activities in neurons (stimulation of choline acetyltransferase [143] and of neurite outgrowth [144]). Most prominently, however, FGF has the ability to prolong the survival of various central and peripheral neurons in culture [81, 143, 144] and in vivo [145–147]. These results suggest that FGF may be a neurotrophic factor. Since FGF is produced in neurons, one should consider that some of its effects on those cells may be under autocrine control.

Other Activities of FGF. Basic FGF modulates pituitary endocrine activity by potentiating selectively the releasing factor-induced secretion of certain pituitary hormones (TSH, prolactin) [148]. Although the mitogen is synthesized in large quantities by the pituitary (in the follicle cells [149]) and is released from cultured pituitary cells [150], bFGF does not appear to be secreted into the blood stream like typical pituitary hormones since no bFGF can be found in the circulation. FGF also modulates steroid hormone production; it suppresses FSH-induced aromatase activity [151] and LH receptor expression [152] in granulosa cells. These data suggest possible roles of FGF in pituitary and ovarian hormone physiology.

Still other reported FGF activities include various effects on endothelial cells such as the stimulation of (a) prostaglandin [153] and prostacyclin [154], factor VIII-related surface antigen [153], and converting enzyme [155] synthesis; (b) release of neurotensin and somatostatin from a neuropeptide-synthesizing thyroid cell line [156], and (c) translocation of nonhistone proteins to the nucleus [157]. Rather unexpectedly, bFGF can also function as an inhibitor of cell proliferation in certain tumor cells [158]. Not enough information is available to assess the biological significance of those data.

FGF in Disease and Therapy

The expression of FGF activity is certainly under strict regulatory control but it is presently unknown how this control is effected. Do inhibitors restrict FGF activity [159]? Is FGF bound to extracellular matrix components inactive until it is released by regulatory signals? Whatever the mechanisms, it is likely that their disturbance would result in a pathological condition. Experimental evidence suggests that excessive FGF action, or the lack of it, may indeed contribute to disease. Most importantly, FGF appears to play a role in the formation of solid tumors. The earlier hypothesis by Folkman [160], stating that a 'tumor angiogenesis factor' is required for continued tumor growth, seems to be validated by the discovery of FGF. Many tumor cells synthesize FGF and, furthermore, are probably capable of activating the abundant FGF present in most host tissues by degradation of the extracellular matrix of host tissue. FGF has various properties that would support tumor growth: it stimulates the proliferation of tumor cells [35, 43, 45], tumor neovascularization, and the invasive capacity of tumor cells, i.e. their ability to produce plasminogen activator [161]. The involvement of FGF in tumor growth is also supported by the finding that patients with kidney or bladder cancer have elevated urinary FGF levels [50]. Finally, the existence in some tumor tissues of oncogenes which code for FGF-like proteins (see above) must be taken into account. Perhaps the strongest evidence for a role of FGF in tumor pathology comes from an experiment showing that the growth of a chondrosarcoma is significantly retarded in mice treated with anti-bFGF antibodies [100].

FGF has also been implicated in other diseases, e.g. retinopathies which are characterized by excessive vascularization of the retina [47]. The effect of the mitogen on fibroblasts (stimulation of cell proliferation and collagen production) may explain the association of fibrosis with neovascularization in this condition and in retrolental fibroplasia. Furthermore, FGF, by virtue of its ability to stimulate collagenase release [162], may contribute to the pathology of chronic inflammatory conditions such as in rheumatoid arthritis and osteoarthritis which are characterized by cartilage destruction.

FGF with its wide spectrum of mitogenic and nonmitogenic activities on various cell types in many tissues and its ability to induce extracellular matrix production and neovascularization is an obvious candidate for therapeutic application in wound healing. Its potential usefulness has been demonstrated in various animal wound healing models [91, 120–125, 163, 164] involving mostly the skin and the eye. While most skin injuries heal spontaneously, FGF may be beneficial in slow-healing or nonhealing wounds, or in the treatment of burns. Furthermore, the mitogenic and regenerative properties of FGF on cells like fibroblasts, myoblasts, chondrocytes, and osteoblasts may prove beneficial for the accelerated repair of muscle, cartilage and bone tissue. The effects of FGF on neural cells raise the question as to whether FGF might be useful in repairing chronic or acute damage to central or peripheral nervous tissue [145]. The angiogenic properties of FGF might help prevent ischemic damage to the myocardium by accelerating neovascularization around partially blocked coronary arteries.

Macrophage-Derived Fibroblast Growth Factor

Mononuclear phagocytes synthesize and secrete growth-stimulating activity for a number of important tissue cells. This activity has in the past been most often referred to as monocyte-macrophage-derived growth factor (MDGF). MDGF has been described as a mitogen for fibroblasts [165–168] but was also shown to stimulate the proliferation of other cells such as smooth smuscle cells [168], endothelial cells [169–172], chondrocytes and osteoblasts [173], and to be angiogenic [171, 174, 175].

Monocytes and macrophages play important roles in host-defense mechanisms, including inflammation, tissue repair, and the immune response. Macrophages accumulate at sites of tissue damage where they are thought to release substances like MDGF. While MDGF is assumed to aid in the tissue repair process, the possibility has been considered that MDGF also contributes to pathological states. For example, angiogenic activity released by tumor-associated macrophages could stimulate tumor vascularization [172]. Likewise, neovascularization in delayed hypersensitivity reactions of the skin, which coincides in time, distribution and magnitude with macrophage infiltration into the lesioned tissue [176], could be the result of MDGF action. An association between MDGF activity and various lung pathologies is strongly suspected in idiopathic pulmonary fibrosis [177], in lung fibrosis caused by asbestos and other toxic environmental substances [178, 179] and acute lung injury [180]. In those diseases fibroblast growth activity derived from respiratory tract macrophages is thought to contribute to excessive fibroblast proliferation, resulting in fibrosis of the lung tissue.

The chemical nature of MDGF has not been studied in detail. MDGF has so far eluded purification to homogeneity and rigorous chemical characterization. Although MDGF could constitute multiple chemical entities, many of the MDGF activities were thought to be associated with a single as yet unknown protein species [168] which was distinguishable from another known major macrophage product, interleukin-1 [181]. Recent significant advances in the fields of growth factor chemistry and biology now provide new insight into the nature of MDGF. Already early on it was recognized [165] that MDGF possesses properties reminiscent of those of platelet-derived growth factor (PDGF), It is now well established that macrophages produce and secrete PDGF [182, 183]. Moreover, they synthesize bFGF [38], TGF-beta [184], TGF-alpha [185], and TNF [186]. All these factors possess one or the other activity reported for MDGF, and in combination, these mitogens can therefore mimic all activities ascribed to MDGF. In this light it is questionable as to whether the activities so far commonly referred to as MDGF correspond to a unique protein. It is more likely that most of the known MDGF activities are the result of known macrophage cytokines. This is not to say that macrophages could not contain novel growth factors possessing activities resembling those of MDGF. For example, the nonheparin-binding endothelial cell growth activity with an apparent molecular weight of 8–9 kd described by Okabe and Takaku [172] could possibly represent such a molecule. However, only rigorous chemical analysis can establish this.

References

1 Baird A, Esch F, Mormède P, et al: Molecular characterization of fibroblast growth factor: distribution and biological activities in various tissues. Recent Prog Horm Res 1986;42:143–205.

2 Gospodarowicz D, Neufeld G, Schweigerer L: Fibroblast growth factor. Mol Cell Endocrinol 1986;46:187–204.

3 Thomas KA: Fibroblast growth factors. FASEB J 1987;1:434–440.

4 Gospodarowicz D: Localization of a fibroblast growth factor and its effect alone and with hydrocortisone on 3T3 cell growth. Nature 1974;249:123–127.

5 Thomas KA, Riley M, Lemmon SK, et al: Brain fibroblast growth factor: nonidentity with myelin basic protein fragments. J Biol Chem 1980;255:5517–5520.

6 Thomas KA, Rios-Candelore M, Fitzpatrick S: Purification and characterization of acidic fibroblast growth factor from bovine brain. Proc Natl Acad Sci USA 1984;81:357–361.

7 Böhlen P, Baird A, Esch F, et al: Isolation and partial molecular characterization of pituitary fibroblast growth factor. Proc Natl Acad Sci USA 1984;81:5364–5368.

8 Gimenez-Gallego G, Rodkey K, Bennett C, et al: Brain-derived acidic fibroblast growth factor: complete amino acid sequence and homologies. Science 1985; 230:1385–1387.

9 Esch F, Baird A, Ling N, et al: Primary structure of bovine pituitary fibroblast growth factor (FGF) and comparison with the amino-terminal sequence of bovine acidic FGF. Proc Natl Acad Sci USA 1985;82:6507–6511.

10 Baird A, Esch F, Böhlen P, et al: Isolation and partial characterization of an endothelial cell growth factor from bovine kidney: Homology with basic fibroblast growth factor. Regul Peptides 1985;12:201–214.

11 Gospodarowicz D, Cheng J, Lui GM, et al: Corpus luteum angiogenic factor is related to fibroblast growth factor. Endocrinology 1985;117:2383–2391.

12 Gospodarowicz D, Baird A, Cheng J, et al: Isolation of fibroblast growth factor from bovine adrenal gland: physicochemical and biological characterization. Endocrinology 1986;118:82–90.
13 Ueno N, Baird A, Esch F, et al: Purification and partial characterization of a mitogenic factor from bovine liver: structural homology with basic fibroblast growth factor. Regul Peptides 1986;16:135–145.
14 Ueno N, Baird A, Esch F, et al: Isolation of an amino terminal extended form of basic fibroblast growth factor. Biochem Biophys Res Commun 1986;138:580–588.
15 Gautschi P, Fràter-Schroeder M, Müller T, et al: Chemical and biological characterization of a truncated form of acidic fibroblast growth factor from bovine brain. Eur J Biochem 1986;160:357–361.
16 Burgess WH, Mehlman T, Marshak DR, et al: Structural evidence that endothelial cell growth factor beta is the precursor of both endothelial cell growth factor alpha and acidic fibroblast growth factor. Proc Natl Acad Sci USA 1986;83:7216–7220.
17 Gautschi-Sova P, Fràter-Schröder M, Jiang ZP, et al: Acidic fibroblast growth factor is present in non-neural tissue: Isolation and chemical characterization from bovine kidney. Biochemistry 1987;26:5844–5847.
18 Bertolini J, Hearn MTW: Isolation, characterization and tissue localization of an N-terminal-truncated variant of fibroblast growth factor. Mol Cell Endocrinol 1987; 51:187–199.
19 Simpson RJ, Moritz RL, Lloyd CJ, et al: Primary structure of ovine pituitary basic fibroblast growth factor. FEBS Lett 1987;224:128–132.
20 Abraham JA, Whang JL, Tumolo A, et al: Human basic fibroblast growth factor: nucleotide sequence and genomic organization. Eur Mol Biol Org J 1986;5:2523–2528.
21 Gautschi-Sova P, Müller T, Böhlen P: Amino acid sequence of human acidic fibroblast growth factor. Biochem Biophys Res Commun 1986;140:874–880.
22 Jaye M, Howk R, Burgess W, et al: Human endothelial cell growth factor: cloning, nucleotide sequence, and chromosome localization. Science 1986;233:541–545.
23 Abraham JA, Mergia A, Whang JL, et al: Nucleotide sequence of a bovine clone encoding the angiogenic protein, basic fibroblast growth factor. Science 1986; 233:545–548.
24 Mergia A, Eddy R, Abraham JA, et al: The genes for basic and acidic fibroblast growth factors are on different human chromosomes. Biochem Biophys Res Commun 1986;138:644–651.
25 Sommer A, Brewer MT, Thompson RC, et al: A form of human basic fibroblast growth factor with an extended amino terminus. Biochem Biophys Res Commun 1987;144:543–550.
26 Story MT, Esch F, Shimasaki S, et al: Amino-terminal sequence of a large form of basic fibroblast growth factor isolated from human benign prostatic hyperplastic tissue. Biochem Biophys Res Commun 1987;142:702–709.
27 Vlodavsky I, Fridman R, Sullivan R, et al: Aortic endothelial cells synthesize basic fibroblast growth factor which remains cell associated and platelet-derived growth factor-like protein which is secreted. J Cell Physiol 1987;131:402–408.
28 Böhlen P, Baird A, Esch F, et al: Acidic fibroblast growth factor (FGF) from bovine brain. Eur Mol Biol Org J 1985;4:1951–1956.
29 Gospodarowicz D, Cheng J, Lui GM, et al: Isolation of brain fibroblast growth factor by heparin-Sepharose affinity chromatography: identity with pituitary fibroblast growth factor. Proc Natl Acad Sci USA 1984;81:6963–6967.

30 Baird A, Esch F, Gospodarowicz D, et al: Retina- and eye-derived endothelial cell growth factors: Partial molecular characterization and identity with acidic and basic fibroblast growth factors. Biochemistry 1985;24:7855–7860.
31 Thompson RW, Wadzinski MG, Sasse J, et al: Isolation of heparin-binding endothelial cell mitogens from normal human myocardium. J Cell Biol 1986;103:300A.
32 Hauschka PV, Mavrakos AE, Iafrati MD, et al: Growth factors in bone matrix. Isolation of multiple types by affinity chromatography on heparin-Sepharose. J Biol Chem 1986;261:12665–12674.
33 Schweigerer L, Neufeld G, Friedman J, et al: Capillary endothelial cells express basic fibroblast growth factor, a mitogen that promotes their own growth. Nature 1987; 325:257–259.
34 Schweigerer L, Ferrara N, Haaparanta T, et al: Basic fibroblast growth factor: expression in cultured cells derived from corneal endothelium and lens epithelium. Exp Eye Res 1988;46:71–80.
35 Moscatelli D, Presta M, Joseph-Silverstein J, et al: Both normal and tumor cells produce basic fibroblast growth factor. J Cell Physiol 1986;129:273–276.
36 Schweigerer L, Neufeld G, Friedman J, et al: Basic fibroblast growth factor: production and growth stimulation in cultured adrenal cortex cells. Endocrinology 1987;120:796–800.
37 Neufeld G, Ferrara N, Schweigerer L, et al: Bovine granulosa cells produce basic fibroblast growth factor. Endocrinology 1987;121:597–603.
38 Baird A, Mormède P, Böhlen P: Immunoreactive fibroblast growth factor in cells of peritoneal exudate suggests its identity with macrophage-derived growth factor. Biochem Biophys Res Commun 1985;126:358–364.
39 Winkles JA, Friesel R, Burgess WH, et al: Human vascular smooth muscle cells both express and respond to heparin-binding growth factor I (endothelial cell growth factor). Proc Natl Acad Sci USA 1987;84:7124–7128.
40 Huang SS, Tsai CC, Adams SP, et al: Neuron localization and neuroblastoma cell expression of brain-derived growth factor. Biochem Biophys Res Commun 1987; 144:81–87.
41 Lobb R, Sasse J, Sullivan R, et al: Purification and characterization of heparin-binding endothelial cell growth factors. J Biol Chem 1986;261:1924–1928.
42 Klagsbrun M, Sasse J, Sullivan R, et al: Human tumor cells synthesize an endothelial cell growth factor that is structurally related to basic fibroblast growth factor. Proc Natl Acad Sci USA 1986;83:2448–2452.
43 Schweigerer L, Neufeld G, Mergia A, et al: Basic fibroblast growth factor in human rhabdomyosarcoma cells: Implications for the proliferation and neovascularization of myoblast-derived tumors. Proc Natl Acad Sci USA 1987;84:842–846.
44 Schweigerer L, Neufeld G, Gospodarowicz D: Basic fibroblast growth factor is present in cultured human retinoblastoma cells. Invest Ophthalmol Vis Sci 1987;28:1838–1843.
45 Libermann TA, Friesel R, Jaye M, et al: An angiogenic growth factor is expressed in human glioma cells. Eur Mol Biol Org J 1987;6:1627–1632.
46 Mascarelli F, Raulais D, Counis MF, et al: Characterization of acidic and basic fibroblast growth factors in brain, retina and vitreous chick embryo. Biochem Biophys Res Commun 1987;146:478–486.
47 Baird A, Culler F, Jones KL, et al: Angiogenic factor in human ocular fluid. Lancet 1985;ii:563.

48 Hamerman D, Taylor S, Kirschenbaum I, et al: Growth factors with heparin binding affinity in human synovial fluid. Proc Soc Exp Biol Med 1987;186:384–389.
49 Chodak GW, Shing Y, Borge M, et al: Presence of heparin binding growth factor in mouse bladder tumors and urine from mice with bladder cancer. Cancer Res 1986;46:5507–5510.
50 Chodak GW, Hospelhom V, Judge SM, et al: Increased levels of fibroblast growth factor-like activity in urine from patients with bladder or kidney cancer. Cancer Res 1988;48:2083–2088.
51 Shing Y, Folkman J, Sullivan R, et al: Heparin affinity: purification of a tumor-derived capillary endothelial cell growth factor. Science 1984;223:1296–1299.
52 Vlodavsky I, Folkman J, Sullivan R, et al: Endothelial cell-derived basic fibroblast growth factor: Synthesis and deposition into subendothelial extracellular matrix. Proc Natl Acad Sci USA 1987;84:2292–2296.
53 Folkman J, Klagsbrun M, Sasse J, et al: A heparin-binding angiogenic protein-basic fibroblast growth factor is stored within basement membrane. Am J Pathol 1988;130:393–400.
54 Baird A, Ling N: Fibroblast growth factors are present in the extracellular matrix produced by endothelial cells in vitro: Implication for a role of heparinase-like enzymes in the neovascular response. Biochem Biophys Res Commun 1987; 142:428–435.
55 Neufeld G, Gospodarowicz D: The identification and partial characterization of the fibroblast growth factor receptor of baby hamster kidney cells. J Biol Chem 1985; 260:13860–13868.
56 Friesel R, Burgess WH, Mehlman T, et al: The characterization of the receptor for endothelial cell growth factor by covalent ligand attachment. J Biol Chem 1986; 261:7581–7584.
57 Olwin BB, Hauschka SD: Identification of the fibroblast growth factor receptor of Swiss 3T3 cells and mouse skeletal muscle myoblasts. Biochemistry 1986;25:3487–3492.
58 Moenner M, Chevallier B, Badet J, et al: Evidence and characterization of the receptor to eye-derived growth factor I, the retinal form of basic fibroblast growth factor, on bovine epithelial lens cells. Proc Natl Acad Sci USA 1986;83:5024–5028.
59 Neufeld G, Gospodarowicz D, Dodge L, et al: Heparin modulation of the neurotrophic effects of acidic and basic fibroblast growth factors, and nerve growth factor on PC12 cells. J Cell Physiol 1987;131:131–140.
60 Moscatelli D, Presta M, Joseph-Silverstein J, et al: Presence of basic fibroblast growth factor in a variety of cells and its binding to cells; in Rifkin DB, Klagsbrun M (eds): Angiogenesis (Current Communications in Molecular Biology). Cold Spring Harbor, Cold Spring Harbor Laboratory Press, 1987, pp 47–51.
61 Moscatelli D: High and low affinity binding sites for basic fibroblast growth factor on cultured cells: Absence of a role for low affinity binding in the stimulation of plasminogen activator production by bovine capillary endothelial cells. J Cell Physiol 1987;131:123–130.
62 Gospodarowicz D, Massoglia S, Cheng J, et al: Isolation of bovine fibroblast growth factor purified by fast protein liquid chromatography (FPLC). Partial chemical and biological characterization. J Cell Physiol 1985;122:323–332.
63 Westphal M, Brunken M, Rohde E, et al: Growth factors in cultured glioma cells: differential effects of FGF, EGF, and PDGF. Cancer Lett 1988;38:283–296.

64 Presta M, Moscatelli D, Joseph-Silverstein J, et al: Purification from a human hepatoma cell line of a basic fibroblast growth factor-like molecule that stimulates capillary endothelial cell plasminogen activator production, DNA synthesis, and migration. Mol Cell Biol 1986;6:4060–4066.
65 Halaban R, Ghosh S, Baird A: bFGF is the putative natural growth factor for human melanocytes. In Vitro Cell Dev Biol 1987:23:47–52.
66 Muller R, Bravo R, Burckhardt J: Induction of c-fos gene and protein by growth factors precedes activation of c-myc. Nature 1984;312:716–720.
67 Tsuda T, Hamamori Y, Yamashita T, et al: Involvement of three intracellular messenger systems, protein kinase C, calcium ion and cyclic AMP, in the regulation of c-fos gene. FEBS Lett 1986;208:39–42.
68 Magnaldo I, L'Allemain G, Chambard JC, et al: The mitogenic signaling pathway of fibroblast growth factor is not mediated through polyphosphoinositide hydrolysis and protein kinase C activation in hamster fibroblasts. J Biol Chem 1986; 261:16916–16922.
69 Mioh H, Chen JK: Acidic heparin binding growth factor transiently activates adenylate cyclase activity in human adult arterial smooth muscle cells. Biochem Biophys Res Commun 1987;146:771–776.
70 Tsuda T, Kaibuchi K, Kawahara Y, et al: Induction of protein kinase C activation and Ca^{2+} mobilization by fibroblast growth factor in Swiss 3T3 cells. FEBS Lett 1985;191:205–210.
71 Takeyama Y, Tanimoto T, Hoshijima M, et al: Enhancement of fibroblast growth factor-induced diacylglycerol formation and protein kinase C activation by colon tumor-promoting bile acid in Swiss 3T3 cells. Different modes of action between bile acid and phorbol ester. FEBS Lett 1986;197:339–343.
72 Huang SS, Huang JS: Association of bovine brain-derived growth factor receptor with protein tyrosine kinase activity. J Biol Chem 1986;261:9568–9571.
73 Chambard JC, Paris S, L'Allemain G, et al: Two growth factor signalling pathways in fibroblasts distinguished by pertussis toxin. Nature 1987;326:800–803.
74 Moscat J, Moreno F, Herrero C, et al: Endothelial cell growth factor and ionophore A23187 stimulation of production of inositol phosphates in porcine aorta endothelial cells. Proc Natl Acad Sci USA 1988;85:659–663.
75 Blanquet PR, Paillard S, Courtois Y: Influence of fibroblast growth factor on phosphorylation and activity of a 34 kd lipocortin-like protein in bovine epithelial lens. FEBS Lett 1988;229:183–187.
76 Pelech S, Olwin BB, Krebs EG: Fibroblast growth factor treatment of Swiss 3T3 cells activates a subunit S6 kinase that phosphorylates a synthetic peptide substrate. Proc Natl Acad Sci USA 1986;83:5968–5972.
77 Togari A, Dickens G, Kuzuya H, et al: The effect of fibroblast growth factor on PC12 cells. J Neurosci 1985;5:307–316.
78 Hamilton RT, Nilsen-Hamilton M, Adams GA: Superinduction by cycloheximide of mitogen-induced secreted proteins produced by Balb/c 3T3 cells. J Cell Physiol 1985;123:201–208.
79 Rizzino A, Ruff E: Fibroblast growth factor induces the soft agar growth of two non-transformed cell lines, In Vitro Cell Dev Biol 1986;22:749–755.
80 Baird A, Schubert D, Ling N, et al: Receptor and heparin-binding domains of basic fibroblast growth factor. Proc Natl Acad Sci USA 1988;85:2324–2328.

81 Schubert D, Ling N, Baird A: Multiple influences of a heparin-binding growth factor on neuronal development. J Cell Biol 1987;104:635–643.
82 Seno M, Sasada R, Iwane M. et al: Stabilizing basic fibroblast growth factor using protein engineering. Biochem Biophys Res Commun 1988;151:701–708.
83 McKeehan WL, Crabb JW: Isolation and characterization of different molecular and chromatographic forms of heparin-binding growth factor 1 from bovine brain. Anal Biochem 1987;164:563–569.
84 Gospodarowicz D, Cheng J: Heparin protects basic and acidic FGF from inactivation. J Cell Physiol 1986;128:475–484.
85 Dickson C, Peters G: Potential oncogene product related to growth factors. Nature 1987;326:833.
86 Delli Bovi P, Curatola AM, Kern FG, et al: An oncogene isolated by transfection of Kaposi's sarcoma DNA encodes a growth factor that is a member of the FGF family. Cell 1987;50:729–737.
87 Yoshida T, Miyagawa K, Odagiri H, et al: Genomic sequence of hst, a transforming gene encoding a protein homologous to fibroblast growth factors and the int-2-encoded protein. Proc Natl Acad Sci USA 1987;84:7305–7309.
88 Thomas KA, Rios-Candelore M, Gimenez-Gallego G, et al: Pure brain-derived acidic fibroblast growth factor is a potent angiogenic vascular endothelial cell mitogen with sequence homology to interleukin-1. Proc Natl Acad Sci USA 1985;82:6409–6413.
89 Gospodarowicz D, Moran J, Braun D, et al: Clonal growth of bovine vascular endothelial cells: fibroblast growth factor as a survival agent. Proc Natl Acad Sci USA 1976;73:4120–4124.
90 Maciag T, Hoover GA, Weinstein R: High and low molecular weight forms of endothelial cell growth factor. J Biol Chem 1982;257:5333–5336.
91 Buntrock P, Buntrock M, Marx I, et al: Stimulation of wound healing, using brain extract with fibroblast growth factor activity. III. Electron microscopy, autoradiography, and ultrastructural autoradiography of granulation tissue. Exp Pathol 1984; 26:247–254.
92 Terranova VP, DiFlorio R, Lyall RM, et al: Human endothelial cells are chemotactic to endothelial cell growth factor and heparin. J Cell Biol 1985;101:2330–2334.
93 Vlodavsky I, Johnson LK, Greenburg G, et al: Vascular endothelial cells maintained in the absence of fibroblast growth factor undergo structural and functional alterations that are incompatible with their in vivo differentiated properties. J Cell Biol 1979;83:468–486.
94 Tseng G, Savion N, Stern R, et al: Fibroblast growth factor modulates synthesis of collagen in cultured vascular endothelial cells. Eur J Biochem 1982;122:355–360.
95 Montesano R, Vasalli JD, Baird AS, et al: Basic fibroblast growth factor induces angiogenesis in vitro. Proc Natl Acad Sci USA 1986;83:7297–7301.
96 Jaye M, McConathy E, Drohan W, et al: Modulation of the sis gene transcript during endothelial cell differentiation in vitro. Science 1985;228:882–885.
97 Risau W: Developing brain produces an angiogenesis factor. Proc Natl Acad Sci USA 1986;83:3855–3859.
98 Risau W, Ekblom P: Production of a heparin-binding angiogenesis factor by the embryonic kidney. J Cell Biol 1986;103:1101–1107.
99 Risau W, Gautschi-Sova P, Böhlen P: Endothelial cell growth factors in embryonic and adult chick brain are related to human acidic fibroblast growth factor. Eur Mol Biol Org J 1988;7:959–962.

100 Baird A, Mormède P, Böhlen P: Immunoreactive fibroblast growth factor (FGF) in a transplantable chondrosarcoma: Inhibition of tumor growth by antibodies to FGF. J Cell Biochem 1986;30:79–85.
101 Bassett DL: The changes in the vascular pattern of the ovary of the albino rat during the estrous cycle. Am J Anat 1943;73:251–262.
102 Jakob W, Jentzsch B, Bauersberger B, et al: Demonstration of angiogenesis activity in corpus luteum of cattle. Exp Pathol 1977;13:231–242.
103 Gospodarowicz D, Thakral TK: Production of a corpus luteum angiogenic factor responsible for proliferation of capillaries and neovascularization of the corpus luteum. Proc Natl Acad Sci USA 1978;75:847–851.
104 Koos RD, Lemaire WJ: Evidence for an angiogenic factor from rat follicles: in Greenwald GS, Terranova PF (eds): Factors Regulating Ovarian Function. New York, Raven Press, 1983, pp 191–195.
105 Frederick HL, Shimanuki T, DiZerega GS: Initiation of angiogenesis by human follicular fluid. Science 1984;224:389–390.
106 Schreiber AB, Winkler ME, Derynck R: Transforming growth factor-alpha: A more potent angiogenic mediator than epidermal growth factor. Science 1986;232:1250–1253.
107 Roberts AB, Spoorn MB, Assoian RK, et al: Transforming growth factor type beta: Rapid induction of fibrosis and angiogenesis in vivo and stimulation of collagen formation in vitro. Proc Natl Acad Sci USA 1986;83:4167–4171.
108 Fràter-Schroeder M, Risau W, Hallmann R, et al: Tumor necrosis factor-alpha, a potent inhibitor of endothelial cell proliferation in vitro, is angogenic in vivo. Proc Natl Acad Sci USA 1987;84:5277–5281.
109 Fett J, Strydom D, Lobb R, et al: Isolation and characterization of angiogenin, an angiogenic protein from human carcinoma cells. Biochemistry 1985;24:5480–5486.
110 Folkman J, Klagsbrun M: Angiogenic factors. Science 1987;235:442–447.
111 Slack J, Darlington B, Heath H, et al: Mesoderm induction in early Xenopus embryos by heparin-binding growth factors. Nature 1987;326:197–200.
112 Kimelman D, Kirschner M: Synergistic induction of mesoderm by FGF and TGF-beta and the identification of an mRNA coding for FGF in the early Xenopus embryo. Cell 1987;51:869–877.
113 Broad TE, Ham RG: Growth and adipose differentiation of sheep preadipocyte fibroblasts in serum-free medium. Eur J Biochem 1983;135:33–39.
114 Rydel RE, Greene LA: Acidic and basic fibroblast growth factors promote stable neurite outgrowth and neuronal differentiation in cultures of PC12 cells. J Neurosci 1987;7:3639–3653.
115 Hamerman D, Sasse J, Klagsbrun M: A cartilage-derived growth factor enhances hyaluronate synthesis and diminishes sulfated glycosaminoglycan synthesis in chondrocytes. J Cell Physiol 1986;127:317–322.
116 Chamberlain CG, McAvoy JW: Evidence that fibroblast growth factor promotes lens fibre differentiation. Curr Eye Res 1987;6:1165–1168.
117 Kardami E, Spector D, Strohman RC: Myogenic growth factor present in skeletal muscle is purified by heparin-affinity chromatography. Proc Natl Acad Sci USA 1985;82:8044–8047.
118 Clegg CH, Linkhart TA, Olwin BB, et al: Growth factor control of skeletal muscle differentiation: commitment to terminal differentiation occurs in G1 phase and is repressed by fibroblast growth factor. J Cell Biol 1987;105:949–956.

119 Wice B, Milbrandt J, Glaser L: Control of muscle differentiation in BC3H1 cells by fibroblast growth factor and vanadate. J Biol Chem 1987;262:1810–1817.
120 Davidson JM, Klagsbrun M, Hill KE, et al: Accelerated wound repair, cell proliferation, and collagen accumulation are produced by a cartilage-derived growth factor. J Cell Biol 1985;100:1219–1227.
121 Fredj-Reygrobellet D, Plouet J, Delayre T, et al: Effects of aFGF and bFGF on wound healing in rabbit corneas. Curr Eye Res 1987;6:1025–1029.
122 Petroutsos G, Courty J, Guimaraes R, et al: Comparison of the effects of EGF, pFGF, and EDGF on corneal epithelium wound healing. Curr Eye Res 1984;3:593–598.
123 Muller G, Courty J, Courtois Y, et al: Utilisation du facteur de croissance dérivé de l'oeil dans le traitement des ulcères de cornée. J Fr Ophthalmol 1985;8:187–192.
124 Gospodarowicz D, Greenburg G: The effects of epidermal and fibroblast growth factors on the repair of corneal endothelial wounds in bovine corneas maintained in organ culture. Exp Eye Res 1979;28:147–157.
125 Jentzsch KD, Wellmitz G, Heder G, et al: A bovine brain fraction with fibroblast growth factor activity inducing articular cartilage regeneration in vivo. Acta Biol Med Germ 1980;39:967–971.
126 Rodan SB, Wesolowski G, Thomas K, et al: Growth stimulation of rat calvaria osteoblastic cells by acidic fibroblast growth factor. Endocrinology 1987;121:1917–1923.
127 Fourtanier AY, Courty J, Muller E, et al: Eye-derived growth factor isolated from bovine retina and used for epidermal wound healing in vivo. J Invest Dermatol 1986;87:76–80.
128 Chua CC, Barritault D, Geiman DE, et al: Induction and suppression of type I collagenase in cultured human cells. Collagen Relat Res 1987;7:277–284.
129 Saksela O, Moscatelli D, Rifkin DB: The opposing effect of basic fibroblast growth factor and transforming growth factor beta on the regulation of plasminogen activator activity in capillary endothelial cells. J Cell Biol 1987;105:957–963.
130 Canalis E, Lorenzo J, Burgess WH, et al: Effects of endothelial cell growth factor on bone remodelling in vitro. J Clin Invest 1987;79:52–58.
131 Sprugel KH, McPherson JM, Clowes AW, et al: Effects of growth factors in vivo. I. Cell ingrowth into porous subcutaneous chambers. Am J Pathol 1987;129:601–613.
132 Turner DC, Bagnaras JT: General Endocrinology, ed 5. Philadelphia, Saunders, 1981, pp 349–386.
133 Gospodarowicz D, Mescher AL: Fibroblast growth factor and vertebrate regeneration; in Riccardi VM, Mulvihill JJ (eds): Advances in Neurology: Neurofibromatosis. New York, Raven Press, 1981, vol 29, pp 149–171.
134 Mescher AL, Gospodarowicz D: Mitogenic effect of a growth factor derived from myelin of denervated regenerates of newt forelimbs. J Exp Zool 1979;207:497–503.
135 Cuny R, Jeanny JC, Courtois Y: Lens regeneration from cultured newt irises stimulated by retina-derived growth factors (EDGFs). Differentiation 1986;32:221–229.
136 Mescher AL, Loh JJ: Newt forelimb regeneration of blastemas in vitro. Cellular response to explantation and effects of various growth-promoting substances. J Exp Zool 1980;216:235–245.
137 Arruti C, Cirillo A, Courtois Y: An eye-derived growth factor regulates epithelial cell proliferation in the cultured lens. Differentiation 1985;28:286–290.

138 Saneto RP, DeVellis J: Characterization of cultured rat oligodendrocytes proliferating in a serum-free, chemically defined medium. Proc Natl Acad Sci USA 1985; 82:3509–3513.
139 Morrison RS, DeVellis J: Growth of purified astrocytes in a chemically defined medium. Proc Natl Acad Sci USA 1981;78:7205–7209.
140 Senior RM, Huang SS, Griffin GL, et al: Brain-derived growth factor is a chemoattractant for fibroblasts and astroglial cells. Biochem Biophys Res Commun 1986; 141:67–72.
141 Weibel M, Pettmann B, Labourdette G, et al: Morphological and biochemical maturation of rat astroglial cells grown in a chemically defined medium: influence of an astroglial growth factor. Int J Dev Neurosci 1985;3:617–630.
142 Gensburger C, Labourdette G, Sensenbrenner M: Brain basic fibroblast growth factor stimulates the proliferation of rat neuronal precursor cells in vitro. FEBS Lett 1987;217:1–5.
143 Unsicker K, Reichert-Preibsch H, Schmidt R, et al: Astroglial and fibroblast growth factors have neurotrophic functions for cultured peripheral and central nervous system neurons. Proc Natl Acad Sci USA 1987;84:5459–5463.
144 Walicke P, Cowan WM, Ueno N, et al: Fibroblast growth factor promotes survival of dissociated hippocampal neurons and enhances neurite extension. Proc Natl Acad Sci USA 1986;83:3012–3016.
145 Baird A, Ueno N, Esch F, et al: Distribution of fibroblast growth factors (FGFs) in tissues and structure-function studies with synthetic fragments of basic FGF. J Cell Physiol 1987;5 (suppl):101–106.
146 Anderson KJ, Dam D, Lee S, et al: Basic fibroblast growth factor prevents death of lesioned cholinergic neurons in vivo. Nature 1988;332:360–361.
147 Otto D, Unsicker K, Grothe C: Pharmacological effects of nerve growth factor and fibroblast growth factor applied to the transsectioned sciatic nerve on neuron death in adult rat dorsal root ganglia. Neurosci Lett 1987;83:156–160.
148 Baird A, Mormede P, Wehrenberg WB: A non-mitogenic pituitary function of fibroblast growth factor: regulation of thyrotropin and prolactin secretion. Proc Natl Acad Sci USA 1985;82:5545–5549.
149 Ferrara N, Schweigerer L, Neufeld G, et al: Pituitary follicular cells produce basic fibroblast growth factor. Proc Natl Acad Sci USA 1987;84:5773–5777.
150 Baird A, Böhlen P, Ling N, et al: Radioimmunoassay for fibroblast growth factor (FGF): release by the bovine anterior pituitary in vitro. Regul Peptides 1985;10:309–317.
151 Adashi EY, Resnick CE, Croft CS, et al: Basic fibroblast growth factor as a regulator of ovarian granulosa cell differentiation: a novel non-mitogenic role. Mol Cell Endocrinol 1988;55:7–14.
152 Baird A, Hsueh AJW: Fibroblast growth factor as an intraovarian hormone: differential regulation of steroidogenesis by an angiogenic factor. Regul Peptides 1986; 16:243–250.
153 Gospodarowicz D, Vlodavsky I, Savion N, et al: Control of the proliferation and differentiation of vascular endothelial cells by fibroblast growth factor; in Bloom F (ed): Peptides: Integrators of Cell and Tissue Function. New York, Raven Press, 1980, pp 1–37.
154 Sebag J, McMeel W: Diabetic retinopathy. Pathogenesis and the role of retina-derived growth factor in angiogenesis. Surv Ophthalmol 1986;30:377–384.

155 Okabe T, Yamagata K, Fujisawa M, et al: Induction by fibroblast growth factor of angiotensin converting enzyme in vascular endothelial cells in vitro. Biochem Biophys Res Commun 1987;145:1211–1216.
156 Zeytin FN, Delellis R: The neuropeptide-synthesizing rat 44-2C cell line: regulation of peptide synthesis, secretion, 3,5′-cyclic adenosine monophosphate efflux, and adenylate cyclase activation. Endocrinology 1987;121:352–360.
157 Polet H: Effects of fibroblastic growth factor on protein degradation, the migration of non-histone proteins to the nucleus and DNA synthesis in diploid fibroblasts. Exp Cell Res 1987;169:178–190.
158 Schweigerer L, Neufeld G, Gospodarowicz D: Basic fibroblast growth factor as a growth inhibitor for cultured human tumor cells. J Clin Invest 1987;80:1516–1520.
159 Böhlen P, Fràter-Schroeder M, Michel T, et al: Inhibitors of endothelial cell proliferation; in Rifkin DB, Klagsbrun M (eds); Angiogenesis. Current Communications in Molecular Biology. Cold Spring Harbor, Cold Spring Harbor Laboratory, 1987, pp 119–124.
160 Folkman J: Antiangiogenesis: New Concept for therapy. Ann Surg 1972;175:409–416.
161 Mira-y-Lopez R, Joseph-Silverstein J, Rifkin DB, et al: Identification of a pituitary factor responsible for enhancement of plasminogen activator activity in breast tumor cells. Proc Natl Acad Sci USA 1986;83:7780–7784.
162 Phadke K: Fibroblast growth factor enhances the interleukin-1-mediated chondrocytic protease release. Biochem Biophys Res Commun 1987;142:448–453.
163 Buntrock P, Jentzsch KD, Heder G: Stimulation of wound healing, using brain extract with fibroblast growth factor activity. I. Quantitative and biochemical studies into formation of granulation tissue. Exp Pathol 1982;21:46–53.
164 Buntrock P, Jentzsch KD, Heder G: Stimulation of wound healing, using brain extract with fibroblast growth factor activity. 2. Histological and morphometric examination of cells and capillaries. Exp Pathol 1982;21:62–67.
165 Leibovich SJ, Ross R: A macrophage-dependent factor that stimulates the proliferation of fibroblasts in vitro. Am J Pathol 1976;84:501–514.
166 Bitterman PB, Rennard SI, Hunninghake GE, et al: Human alveolar macrophage growth factor for fibroblasts: regulation and partial characterization. J Clin Invest 1982;70:806–822.
167 Dohlman J, Payan D, Goetzl E: Generation of a unique fibroblast-activating factor by human monocytes. Immunology 1984;52:577–584.
168 Gimbrone MA, Martin BM, Baldwin WM, et al: Stimulation of vascular growth by macrophage products; in Nossel HL, Vogel HJ (eds): Pathobiology of the Endothelial Cell. New York, Academic Press, 1982, pp 3–17.
169 Greenburg GB, Hunt TK: The proliferative response in vitro of vascular endothelial and smooth muscle cells exposed to wound fluid and macrophages. J Cell Physiol 1978;97:353–360.
170 Ziats NP, Robertson AL: Effects of peripheral blood monocytes on human vascular cell proliferation. Atherosclerosis 1981;38:401–410.
171 Polverini PJ, Leibovich SJ: Activated macrophages induce vascular proliferation. Lab Invest 1984;51:635–642.
172 Okabe T, Takaku F: A macrophage factor that stimulates the proliferation of vascular endothelial cells. Biochem Biophys Res Commun 1986;134:344–350.

173 Rifas L, Shen V, Mitchell K, et al: Macrophage derived growth factor for osteoblast-like cells and chondrocytes. Proc Natl Acad Sci USA 1984;81:4558–4562.
174 Thakral KK, Goodson WH, Hunt TK: Stimulation of wound blood vessel growth by wound macrophages. J Surg Res 1979;26:430–438.
175 Knighton DR, Hunt TK, Scheuenstuhl H, et al: Oxygen tension regulates the expression of angiogenesis factor by macrophages. Science 1983;221:1283–1285.
176 Polverini PJ, Cotran RS, Sholley MM: Endothelial proliferation in the delayed hypersensitivity reaction: an autoradiographic study. J Immunol 1977;118:529–532.
177 Rennard SI, Bitterman PB, Ozaki T, et al: Colchicine suppresses the release of fibroblast growth factors from alveolar macrophages in vitro. The basis of a possible therapeutic approach to the fibrotic disorders. Am Rev Respir Dis 1988;137:181–185.
178 Lemaire I, Beaudoin H, Masse S, et al: Alveolar macrophage stimulation of lung fibroblast growth in asbestos-induced pulmonary fibrosis. Am J Pathol 1986; 122:205–211.
179 Schoenberger CI, Rennard SI, Bitterman PB, et al: Paraquat-induced pulmonary fibrosis. Role of the alveolitis in modulating the development of fibrosis. Am Rev Respir Dis 1984;129:168–173.
180 Kovacs EJ, Kelley J: Intra-alveolar release of a competence-type growth factor after lung injury. Am Rev Respir Dis 1986;133:68–72.
181 Estes JE, Pledger WJ, Gillespie YG: Macrophage-derived growth factor for fibroblasts and interleukin-1 are distinct entities. J Leukocyte Biol 1984;35:115–129.
182 Shimokado K, Raines EW, Madtes DK, et al: A significant part of macrophage-derived growth factor consists of at least two forms of PDGF. Cell 1985;43:277–286.
183 Mornex JF, Martinet Y, Yamaguchi K, et al: Spontaneous expression of the sis gene and release of a platelet-derived growth factor-like molecule by human alveolar macrophages. J Clin Invest 1986;78:61–66.
184 Assoian RK, Fleudelys BE, Stevenson HC: Expression and secretion of type beta transforming growth factor by activated human macrophages. Proc Natl Acad Sci USA 1987;84:6020–6024.
185 Madtes DK, Raines EW, Sakariassen KS, et al: Induction of transforming growth factor alpha in activated alveolar macrophages. Cell 1988;53:285–293.
186 Austgulen R, Espevik T, Nissen-Meyer J: Fibroblast growth-stimulatory activity released from human monocytes. The contribution of tumour necrosis factor. Scand J Immunol 1987;26:621–629.

Dr. Peter Böhlen, Department of Protein Chemistry,
American Cyanamid Company, Medical Research Division, Lederle Laboratories,
Middletown Road, Pearl River, NY 10965 (USA)

Subject Index